# Na-linked Transport of Organic Solutes

The Coupling between Electrolyte and Nonelectrolyte
Transport in Cells

A Symposium Held under the Auspices of the International Union
of Physiological Sciences and of the Gesellschaft für Biologische Chemie
in Erbach/Rheingau, Germany, August 3–5, 1971
as a Satellite Symposium of the XXV International Congress
of Physiological Sciences, Munich, Germany

Edited by

## Erich Heinz

With the Assistance of H. Kromphardt and B. Pfeiffer

With 118 Figures

Springer-Verlag Berlin Heidelberg New York 1972

Some of the contributions to this volume has been presented in abstract form in Hoppe-Seylers Z. f. Physiol. Chemie **353** (1972).

The symposium was generously supported by Schering A. G., Berlin.

ISBN 3-540-05849-4 Springer-Verlag Berlin Heidelberg New York
ISBN 0-387-05849-4 Springer-Verlag New York Heidelberg Berlin

# Contents

# Opening Remarks

E. Heinz

Ladies and Gentlemen,

I welcome you to this symposium which, I hope, will give us the opportunity of
a personal confrontation between proponents of different views on the main topic,
the relation between electrolytes and transport of organic solutes in cells. Even
if a final agreement can hardly be expected at this meeting, it may least help
to clear away misunderstandings and clarify the issue, so that new and more crucial experiments may emerge to test it. This kind of encounter is the more de-
sirable as the main controversies, as I see them, seem to concern much more
the interpretations of observations than the observations themselves. Before stat-
ing the issues let me first try to ascertain how far we are on common ground:

Most workers agree that in certain cells and tissues  several organic solutes, in
particular sugars and amino acids, are transported actively, i. e. against their
electro-chemical potential gradient, with the expenditure of metabolic energy.
Most workers agree that many of these active transport systems are activated
by extracellular Na ions and, less certainly, by intracellular K ions, and that this
transport is inhibited if the intracellular $Na^+$ and/or the extracellular $K^+$ are increased. Moreover, there seems little doubt that in this system the movement
of the organic solute is as a rule accompanied by the parallel movement of Na
ions and the antiparallel movement of K ions. Many workers believe that this
joint movement between  non-electrolytes and electrolytes occurs by an intrinsic-
ally fixed stoichiometric ratio, even though there is no unanimity as to whether
this intrinsic ratio can be reliably determined by the conventional methods. Finally, most workers seem to agree that cardiac glycosides, like ouabain, inhibit
these transport systems as they do the Na/K pump,  even though there is no agreement as to whether these drugs act on the transport of organic solutes directly
or via the alkali-ion distribution. This seems to me a brief delimitation of the
common ground. Although the detailed views on these basic points vary from
worker to worker, partly owing to tissue and species differences, there does not
seem to be much fundamental or serious disagreement among most of the people
present at this meeting. What are the controversies, then?

Obviously any transport process requires that the uphill movement of the trans-
portee be coupled to an exergonic process in order that the energy released by
the latter be utilized by the former. Two fundamentally different ways of such
coupling are conceivable: First, there may be a direct linkage between the transport process and a chemical reaction. This kind of coupling is often called "chemi-
osmotic", and the resulting transport, "primary" active. Second, the transport
can be thought of being coupled to another flow. This kind of coupling could, in
analogy to the former one, be called "osmo-osmotic". In order that energy be
made available, the flow of the solute to which the transport is coupled has to be
downhill. To the extent that electrochemical potential gradient of downhill-moving

solute is maintained at the expense of metabolic energy, possibly by a direct (chemi-osmotic) coupling, the active transport of the organic solute depends on metabolic energy. Since this dependence is only indirect, the resulting active transport is called "secondary" active. Even though metabolic energy, e. g. ATP, is required in both cases, the immediate driving force for the transport of the organic solute is different: in primary active transport it is the affinity of the chemical reaction, and in secondary transport the electrochemical potential gradient of the downhill-flowing solute, e. g. of Na or K ions. The main question, then, to be argued in this meeting would be: Is the active transport of the organic solutes primary or secondary active, in other words, is it coupled to a chemical reaction, e. g. the hydrolysis of ATP, or coupled to the inward movement of $Na^+$ and the outward movement of $K^+$. This is the main issue of the controversy.

Before the effects of electrolytes on organic solute transport became known, many workers firmly believed that the active transport of amino acids and sugars, wherever it occurs, is primary active. In the last decade the number of primary systems seems to have dwindled considerably. Only a few of them have survived, such as the phospho-transferase system, the Na/K-ATPase system, and possibly the redox pump, wherever it may exist. For most other transport systems, especially for those of organic solutes, the primary nature has largely been challenged and the "gradient hypothesis", which assumes that these solutes are driven by the electro-chemical potential gradients of electrolyte ions, has been given preference by many workers. Accordingly Vidawer and Eddy have demonstrated beyond doubt that metabolically inhibited cells may accumulate glycine, provided that an appropriate electrolyte gradient is present. It is therefore difficult to deny that at least part of the active transport of sugars and amino acids can be driven by these gradients. On the other hand, it has repeatedly been shown that in actively metabolizing cells this transport is much more effective than it is in inhibited cells, even with the same ion gradients present. The basic question, therefore, has to be somewhat modified: it is no longer whether energy for the active transport of organic solutes can be utilized from electrolyte gradients, but whether this energy is sufficient. I am afraid that this question has not been answered satisfactorily up to the present time. Whether a direct metabolic (chemi-osmotic) coupling is involved, even though it can at best account for only a part of the required energy, is still a crucial issue.

Often teleological arguments are used to support the one or the other hypothesis. So it is often maintained that the gradient hypothesis is more economical for the cell because only a single transport system, that of electrolyte ions, is required to transport both electrolytes and organic solutes. This economy, however, is certainly not very impressive as far as energy is concerned. Each of the two transport processes involved is bound to have a limited efficiency so that a substantial loss of energy will be unavoidable in both cases. Thus, it can hardly make much difference energetically  whether the transport of the organic solute is linked directly or indirectly  to the metabolism. On the other hand, the gradient hypothesis may be more economical in the requirement of intricate coupling devices, as an osmo-osmotic coupling appears to be much simpler than a chemi-osmotic coupling. Still more puzzling from a teleological point of view would be the finding that the cell uses two different coupling devices and two energy sources for the same substrate, if the transport of the amino acid or the sugar were both primary and secondary active. One possible answer could be that the primary transport is the predominant one and that the coupling to the electrolyte gradients is merely accidental. But before speculating on this question it ought to be established first whether primary transport occurs here at all. Anyway, teleological considerations seem to be of little help in this context.

The present meeting will be run in a somewhat unorthodox fashion. Anticipating that most controversies concern interpretations rather than the above-mentioned experimental observations, we shall give comparatively little time to the mere presentation of new experimental findings, but much more to focussed discussions on specific topics. I hope that the chairmen of these special sessions will be strict enough to keep the discussion under control and in line with the scheduled topic. Their sessions should be run like hearings in a trial, where the opponents present their arguments and where other investigators are called upon as witnesses to testify in favor of the one or the other view. It would be splendid if we eventually came up with either a verdict or an agreement, but I doubt whether this will happen.

# Ion Gradient Hypotheses and the Energy Requirement for Active Transport of Amino Acids

John A. Jacquez
Department of Physiology, University of Michigan, Ann Arbor, Michigan 48104
USA

## Introduction

### The evidence for ion cotransports

There is now extensive evidence that active uptake of neutral amino acids depends on the presence of $Na^+$ extracellularly and that there is in fact a cotransport of $Na^+$ and amino acid. There are also some influences of $K^+$ and $H^+$ on such transport although there is little evidence for a direct linkage between movements of $K^+$ and $H^+$ and of amino acids. The work in this area has been reviewed in detail recently by Schultz and Curran (1) so there is no need for me to review the many contributions which have led to the present position. It is important to recall the nature of the evidence for cotransport of $Na^+$; it is primarily of two kinds. First there is the kinetic evidence on the dependence of initial fluxes of amino acids on the extracellular concentration of $Na^+$. Secondly there are direct measurements of the increase in $Na^+$ influx accompanying an amino acid influx. From these come the data on the stoichiometry of the comovement of $Na^+$ and amino acids. Eddy (2) reported an increase in $Na^+$ influx and in $K^+$ efflux accompanying glycine uptake in LS ascites cells, the ratios $\Delta Na^+/\Delta Gly$ and $\Delta K^+/\Delta Gly$ being $0.9 + .1$ and $-0.6 + .1$ respectively. Interestingly, Schafer and Jacquez (3) also found a $\Delta Na^+/\Delta AIB$ of about 1:1 for AIB uptake by Ehrlich ascites cells and Schafer (4) found a $\Delta K^+/\Delta AIB$ of $-0.6$ for this system. Nonetheless the roles of $K^+$ and of $H^+$ are still not clear and in fact the experimental clarification of their roles may be quite difficult because of the following considerations. First of all suppose that the $Na^+$-amino acid cotransport is electrogenic so that there is a net transfer of one positive charge, a $Na^+$ ion, into the cell per carrier cycle. As a result the membrane potential would fall. There is already substantial experimental evidence for a decrease in magnitude of the membrane potential accompanying uptake of sugars and amino acids in a number of cell types (5, 6, 7). Therefore even if there is no directly coupled movement of $K^+$ and $H^+$ the decrease in magnitude of the membrane potential means that the forces tending to give an efflux of $K^+$ and $H^+$ are increased (the membrane potential moves away from the $K^+$ and the $H^+$ equilibrium potentials) and we should expect to see an increase in the efflux of $K^+$ and $H^+$. By the same argument there would be movement of anions into the cell, the relative contributions of the different cations and anions depending on their permeabilities and concentrations. Since for many cells $K^+$ and $Cl^-$ have relatively high permeabilities of approximately the same magnitudes (usually $P_K$ is a little higher than $P_{Cl}$) and are present in high concentrations in comparison with other ions of high permeability, the main effect one would expect is some efflux of $K^+$ and some influx of $Cl^-$, the net charge transfer being almost but not quite enough to neutralize the charge transfer of $Na^+$ by cotransport. Therefore a good test of whether $K^+$ movement is direct or indirect is to measure $\Delta K^+/\Delta AIB$ for a number of different anions in extracellular fluid. If the efflux of $K^+$ is simply a consequence of an electrogenic cotransport of $Na^+$ it should be possible to change the $\Delta K^+/\Delta AIB$ ratio by changing the permeability of the major extracellular anion. With a low permeability anion such as sulfate we should be able to push $\Delta K^+/\Delta AIB$ close to -1. On the other hand, if the $K^+$ efflux is mediated by the amino acid carrier, changing the major anion should not affect the ratio $\Delta K^+/\Delta AIB$. Secondly, intracellular binding of cations would also confuse the picture. The evidence for sodium binding intracellularly and in the nucleus is impressive (8-13). If so, it seems unlikely that there would be sites binding $Na^+$ which would not also bind $K^+$ and $H^+$, albeit with different affinities. Then an increase in intracellular sodium would compete with other cations for the available binding sites and thus increase the intracellular free $K^+$ and $H^+$ and contribute to an efflux of these ions. Thus even if there were no direct coupling of $K^+$ and $H^+$ movement with amino acid movement on a carrier we would expect to find some efflux of $K^+$ and $H^+$ and some influx of $Cl^-$ accompanying amino acid uptake, and we are faced with the much more difficult problem of distinguishing between effluxes directly coupled with the $Na^+$-amino acid cotransport and those which may be secondary to an electrogenic cotransport of $Na^+$ and amino acid.

## Are the ion electrochemical gradients enough?

The demonstration of a cotransport is of course only the beginning. It is important to recognize that a coupling of the movement of two chemical species such as by formation of a ternary complex, NaCS, where C is carrier in a membrane and S represents a substrate, implies that a gradient in one, $Na^+$ or S, will act as a force to move the other. So it is important to determine whether the postulated cotransports adequately explain the concentration gradients obtained. Is the energy obtainable from the sodium electrochemical gradient adequate to explain the concentrative uptake of amino acids found experimentally? Eddy (14), Jacquez and Schafer (15), Potashner and Johnstone (16, 17) and Schafer and Heinz (18) have all reported evidence which suggests that the $Na^+$ electrochemical gradient or the $Na^+$ and $K^+$ gradients combined cannot explain the amino acid concentrations obtained in ascites cells. However there are difficulties in interpreting the data because of the difficulty in determining intracellular activities of the amino acids and ions.

## Approaches to the problem

There are three major approaches to the problem raised in the last section. The first is to manipulate the various forces experimentally and to determine how the amino acid flux depends on the postulated forces. The second is to examine the dependence of the steady-state amino acid concentration gradient on the electrochemical potential gradients in the ions and to calculate whether the energy obtainable from the latter is adequate to explain the amino acid electrochemical gradients. The third is to try to eliminate any possible direct linkage with metabolism with metabolic inhibitors and then check whether amino acids are concentrated to the same extent in such cells as in normal cells that have the same ion gradients.

## Fluxes and forces

In theory this test is rather simple. It asks the question, is the net flux zero when the sum of the forces postulated to act is zero? If so, the forces are adequate to explain the fluxes. In practice the test is not so easy. The major forces which have been implicated are the $Na^+$ electrochemical potential gradient or the sum of the $Na^+$ and $K^+$ electrochemical potential gradients. Fig. 1 and 2 show data published

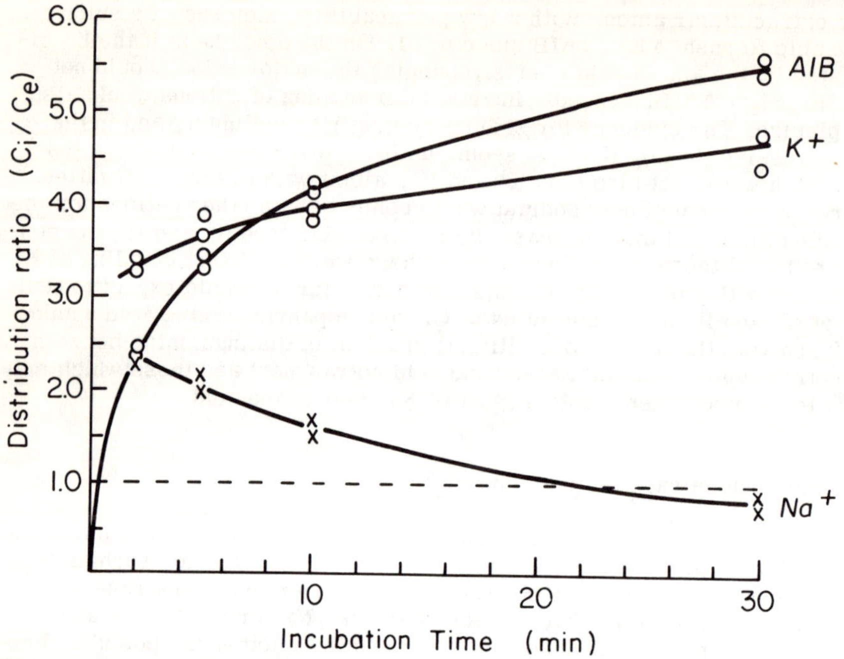



Fig. 1. Time course of uptake of AIB in presence of a reversed $Na^+$ gradient. The initial extracellular contentration was 2 mM. (Reproduced with permission from Jacquez and Schafer, Biochim. Biophys. Acta $\underline{193}$, 268, 1969)

by Jacquez and Schafer (15). From Fig. 1 it appears that AIB is concentrated if the $Na^+$ gradient is reversed and from Fig. 2 it still appears to be concentrated when both the $Na^+$ and the $K^+$ gradients are reversed. But we have evidence for binding of $Na^+$ in the nucleus (8-13). If we assume that the actual intracellular concentration was 1/3 of the measured and that the membrane potential was -12mV then for the 30 minute point in Fig. 1 the $Na^+$ gradient hypothesis adequately explains the concentration ratio of AIB obtained. With the same assumption about the intracellular $Na^+$ one would need a membrane potential of -23mV to account for the AIB distribution ratio at 5 min in Fig. 1. Fig. 2 shows that if both $Na^+$ and $K^+$ gradients are reversed the AIB distribution ratio obtained is less than when only the $Na^+$ gradient is reversed, suggesting that the $K^+$ gradient also plays a role.

Let us suppose that the forces involved are derived from the electrochemical gradients of various solutes. Assume that the fluxes are given as linear combinations of the coupled forces as in irreversible thermodynamics. Then the flux of amino acid, J, is given by equation (1) in which the $X_j$ are the forces.

$$(1) \qquad J = \sum_j L_j X_j$$

Hence if we know the coupling coefficients $L_j$ and vary the $X_j$, a plot of J versus $\sum L_j X_j$ must pass through the origin. Schafer and Heinz (18) have carried out such a test, using the chemical gradients of AIB, $Na^+$ and $K^+$ and have assumed these are directly summable, i. e. all the $L_j$ are the same. This is predicated on the assumption that the stoichiometric coefficients for carrier cotransport for AIB, $Na^+$ and $K^+$ all have the same absolute value and that the cotransports are mandatory.

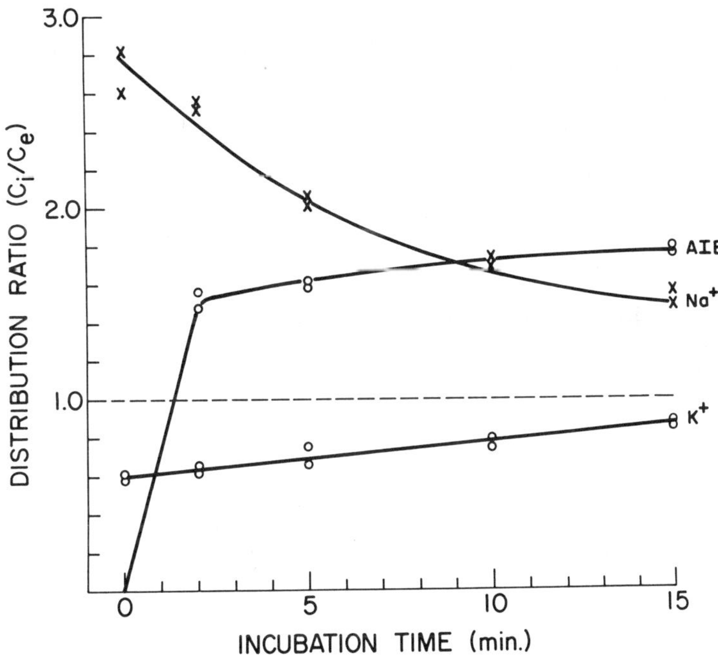

Fig. 2. Time course of uptake of AIB in presence of reversed Na⁺ and K⁺ gradients.
The initial extracellular concentration was 2 mM. (Reproduced with permission
from Jacquez and Schafer, Biochim. Biophys. Acta $\underline{193}$, 368, 1969)

They found that even when the sum of these forces is zero there is a net influx of
AIB into Ehrlich ascites cells.

The above test may be generalized. It is possible that the amino acid flux, J, is
a non-linear function of the forces, $X_j$, as in equation (2).

$$(2) \qquad J = f(X_1, X_2, \ldots)$$

On physical grounds J must be a monotonic function in any one of the forces. If we
choose the sign on any force such that an increase in the force tends to increase
the flux, J is then monotonically non-decreasing in each of the forces. Thus if all
forces but $X_p$ are held constant and $X_p$ is decreased sufficiently, J must decrease.
Furthermore if all of the forces are simultaneously set equal to zero the flux must
become zero.

There are many problems in the interpretation of the flux versus forces test which
should be kept in mind when evaluating the results of such tests. Many cells have
appreciable levels of intracellular amino acids so there is the possibility of a contribution from exchange diffusion in the measurement of initial fluxes. Schafer and
Heinz (18) present evidence to indicate that this is not a significant problem with
AIB uptake in Ehrlich ascites cells. Perhaps the most serious problem is that of
determining the true activities of the ions and amino acids in the cytoplasm. This
is affected both by the binding intracellularly, particularly of the ions, and the usual problem of the activity as a function of the concentration of all solutes. Another
problem, often overlooked, is that there are simultaneous movements of water in

the measurement of fluxes in non-steady states. We can probably discount the effect of this volume flow on amino acid flux through the cross coefficients of irreversible thermodynamics because the reflection coefficients of amino acids must be very close to 1 for most cell membranes. However, this does have an effect if fluxes are reported on the basis of cell water because a water uptake accompanies amino acid uptake and the water flux increases with the amino acid flux. Fluxes on a dry weight basis get around this difficulty. Finally the investigator should be forewarned that it is possible to run initial flux experiments so as to set up gradients in ions which normally serve only to modulate the activity of a transport system in different physiological states but contribute no forces for the transport. To illustrate this I would like to present a hypothetical but not implausible case. In many cells hydrogen ions are in electrochemical equilibrium across the cell membrane; cell $H^+$ is slightly higher than extracellular because of the negative membrane potential. Let us assume the carrier is a protein which has a number of acidic sites, the dissociation of these sites being determined by local $H^+$ concentrations. Thus pH determines the relative concentrations of differently charged carrier species which may not all function equally well in transport. Nonetheless, if under physiological conditions the $H^+$ is in electrochemical equilibrium across the cell membrane, it is thermodynamically impossible for it to contribute any energy to the transport of an amino acid even though $H^+$ does react with the carrier. Now consider how one ordinarily conducts initial flux studies. Usually the cell preparation is washed, warmed to the incubation temperature and then mixed with the amino acid containing medium which is being tested and the initial flux is measured. If the test involves a sudden change in extracellular fluid pH at the start of incubation, the $H^+$ may not initially be in electrochemical equilibrium across the cell membrane and thus contribute a force to the flux of the amino acid. Thus it is possible for the experimenter to introduce forces into his experiments which may not be present in physiological states in vivo and may not ordinarily play any role in driving transport.


## Studies in the steady state

In the steady state the net fluxes of amino acid and of the individual ions are zero. However, there must be a net active transport flux against the electrochemical potential gradient of amino acid which just compensates for the losses by passive diffusion or processes other than those which are a part of the active transport system. This can be used to devise a test for the adequacy of the power supplied by specific ion gradients for maintaining the steady state concentration gradient of an amino acid under specific hypotheses of cotransport. The derivation is simple. Assume that as a result of the operation of the transport system in the steady state the net effect is that k moles of amino acid and m of $Na^+$ are transported into the cell and n of $K^+$ are moved out. It may help to picture this as k molecules of amino acid and m of $Na^+$ binding to the carrier at the outer surface of the cell. The carrier reorients, releases these at the inner surface where it picks up n $K^+$. The carrier again reorients and releases the $K^+$ to the outside. The net effect of a carrier cycle is to transfer k molecules of amino acid from activity $a_e$ to $a_i$ and m ions of $Na^+$ inward and n ions of $K^+$ outwards through their respective electrochemical potential gradients. Let $J_L$ be the leak flux of amino acid due to all sources. The net active transport flux must just compensate for this to maintain a steady state. Thus for this model there is a net carrier transport flux of amino acid of $J_L$ into the cell, $mJ_L/k$ of $Na^+$ into the cell and $nJ_L/k$ of $K^+$ out of the cell. Therefore the rate of energy expenditure needed to maintain the steady state is given by equation (3).

$$(3) \quad P_{AIB} = J_L RT \ln \left(\frac{a_i}{a_e}\right) = \frac{J_L}{k} kRT \ln \left(\frac{a_i}{a_e}\right) = \left(\frac{J_L}{k}\right) RT \ln \left(\frac{a_i}{a_e}\right)^k$$

The maximum rate at which energy could become available from the flux $mJ_L/k$ of $Na^+$ is given by equation (4) and the maximum from the flux of $K^+$ is given by equation (5).

$$(4) \quad P_{Na} = m\left(\frac{J_L}{k}\right) \left[ RT \ln \left(\frac{Na_e}{Na_i}\right) - FV \right]$$

$$(5) \quad P_K = n\left(\frac{J_L}{k}\right) \left[ RT \ln \frac{K_i}{K_e} + FV \right]$$

Here V is the membrane potential, F is the Faraday (23 cal/mole-mV), $Na_e$, $Na_i$ and $K_e$, $K_i$ are the extracellular and intracellular activities of $Na^+$ and $K^+$ respectively. If all of the power from the $Na^+$ and $K^+$ movements is used to move AIB and no other source is available, then equation (6) must hold,

$$(6) \quad RT \ln \left(\frac{a_i}{a_e}\right)^k = RT \ln \left[ \left(\frac{Na_e}{Na_i}\right)^m \left(\frac{K_i}{K_e}\right)^n \right] - (m-n) FV$$

or as in equation (7).

$$(7) \quad \left(\frac{a_i}{a_e}\right)^k = \left(\frac{Na_e}{Na_i}\right)^m \left(\frac{K_i}{K_e}\right)^n e^{-(m-n)FV/RT}$$

If the coupling between the ion movements and the amino acid movement is not 100% efficient then the left side of equation (7) must be less than the right side. This gives us a criterion of adequacy for this specific model.

A number of points should be emphasized about this derivation. There may be other Na and K fluxes, and indeed there must be if this is a true steady state; only the ones coupled to the amino acid transport enter into this derivation. Furthermore, note that the leak flux $J_L$ need not be measured, it cancels out on the two sides of equation (6). Also note the role of the membrane potential; if m = n the transport is electroneutral and the membrane potential does not enter the power calculation but if m $\neq$ n then there is a net transfer of charge per carrier cycle so the membrane potential enters the calculation.

For the sodium gradient hypothesis with k = 1, m = 1 and n = 0 there is an electrogenic cotransport of sodium and our criterion becomes equation (8).

$$(8) \quad \frac{a_i}{a_e} \leq \frac{Na_e}{Na_i} e^{-FV/RT}$$

On the other hand if k = m = n = 1 the transport is electroneutral and the criterion is equation (9).

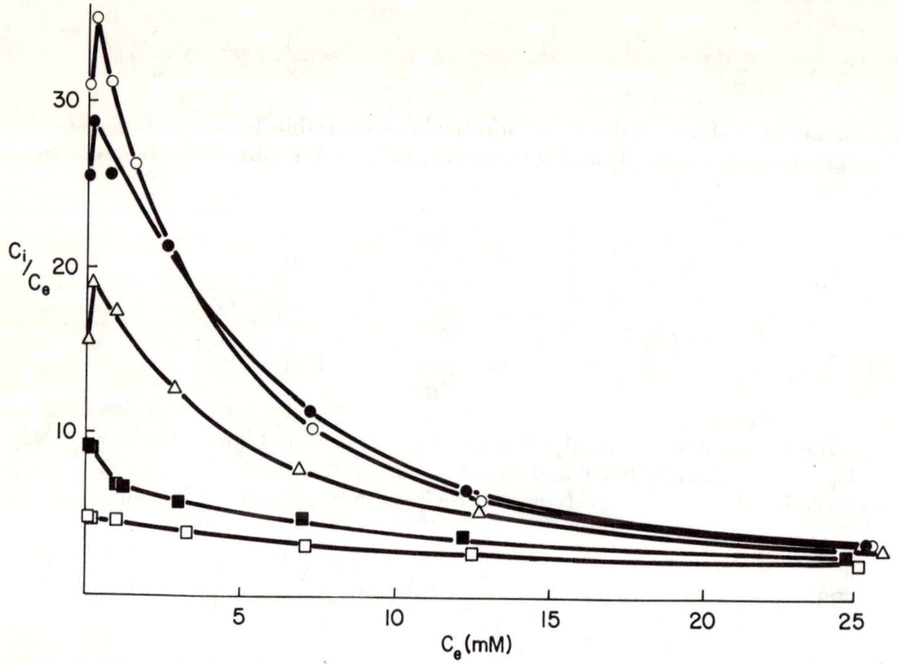

Fig. 3. Variation of $c_i/c_e$ for AIB in Ehrlich ascites cells as function of $c_e$ in the steady state. The incubation medium was a Krebs-Ringer phosphate solution which had the following concentrations of $Na^+$ and $K^+$ in meq./l: -0- Na = 124, K = 22; - ● - Na = 147, K = 2.8; - △ - Na = 40, K = 24; - ■ - Na = 5.0, K = 2.8; - □ - Na = 5.0, K = 24. Choline replaced Na or K.

$$(9) \qquad \frac{a_i}{a_e} \leq \frac{Na_e}{Na_i} \cdot \frac{K_i}{K_e}$$

Such a test does not have all of the problems of the fluxes versus forces test but it is also less specific. It can only eliminate hypotheses which clearly cannot provide the energy input required. Any effect of exchange diffusion can be eliminated by repeated incubation and resuspension of cells in fresh medium so as to wash out any amino acids which contribute by exchange diffusion. The steady-state studies have the further advantage that there are no net movements of water or other solutes. Furthermore the experimenter need not worry about inadvertently setting up gradients in ions which are ordinarily in electrochemical equilibrium across the cell membrane as he should for initial flux studies. However, it shares with the flux versus forces test the same problems about the activities of amino acids and ions intracellularly.

Hence we need to measure the activity ratio of amino acid, $a_i/a_e$, for the steady states. However, we can only measure the concentration ratio, $c_i/c_e$. Fig. 3 shows how $c_i/c_e$ varies with $c_e$ for AIB in Ehrlich ascites tumor cells. These curves are from some recent experiments from my laboratory in which great care was taken to assure steady-state conditions. The peak at about $c_e$ = 0.2 mM was fairly consistently obtained for the high $c_i/c_e$ ratios. Obviously the maximum value of $a_i/a_e$

Fig. 4. Plot of $c_i/c_e$ for AIB in Ehrlich ascites cells as function of $Na_e/Na_i$, in quasi steady-state. (Data from Jacquez and Schafer, Biochim. Biophys. Acta <u>193</u>, 368, 1969)

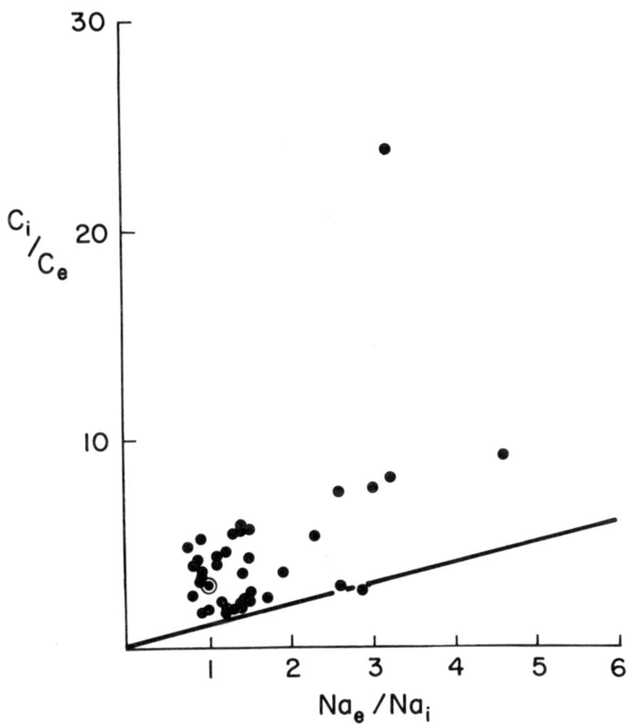


attainable should be used for any test such as that given by equation (7). It is of course possible that binding plays a more important role at lower extracellular concentrations but it probably is not great, although this is again part of the recurrent problem of determining true intracellular activities. The decrease in $c_i/c_e$ as $c_e$ increases is to be expected for a number of reasons. It should appear because of the increasing role of passive leaks in comparison with the saturable transport mechanism as $c_e$ increases, even if the transport is an irreversible mechanism. Superimposed on this may be a loading effect on the transport mechanism if it is in fact reversible. It is difficult for me to believe that the mechanism is not at least in theory reversible. However, there are many examples of enzyme reactions in cells that have their equilibria so far to one side of the reaction that for substrate concentrations available in cells the reactions are practically irreversible. Thus it is not inconceivable that active transport could be "practically" irreversible.

Fig. 4 shows a plot of $c_i/c_e$ for AIB against $Na_e/Na_i$ for the data published by Jacquez and Schafer (13). If we assume a membrane potential of -12mV then $Na_i$ would have to be less than 1/5 the $Na_i$ obtained by dividing total cell sodium by cell water if the point farthest above the $c_i/c_e = Na_e/Na_i$ line is to fit the sodium gradient hypothesis, $k = m$, $n = 0$. These data come from steady-state experiments with ascites cells incubated with ouabain. Fig. 5 shows a plot of $c_i/c_e$ vs $Na_eK_i/Na_iK_e$ for the same experiments and it is obvious that the intracellular Na need only be a little less than the cell water average which was calculated for the model with $k = m = n = 1$ to be adequate. Table I gives results from more recent and more careful steady-state studies at $32^\circ$C and in the absence of ouabain. In this case the maximum $c_i/c_e$ obtained are used in the tests. Obviously the Na-K gradient hypothesis is satisfied by all points, but one would have to assume that the $Na_i$ is as little as

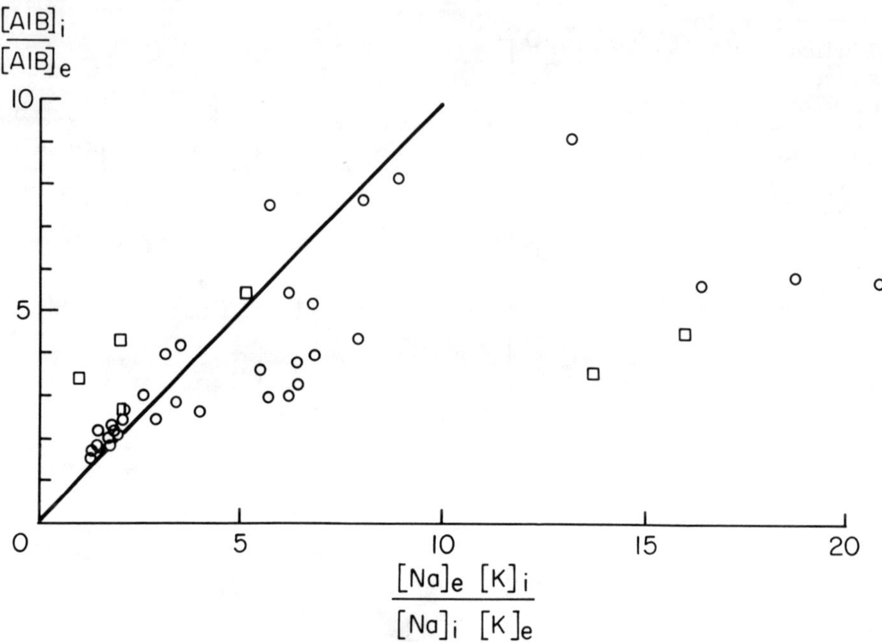

Fig. 5. Plot of quasi-steady state distribution ratio of AIB against $Na_eK_i/Na_iK_e$. (Reproduced with permission from Jacquez and Schafer, Biochim. Biophys. Acta 193, 368, 1969)

17% of the calculated $Na_i$ to make the sodium gradient hypothesis for $k = m = 1$, $n = 0$ fit for a membrane potential of -12mV.

## Excluding a direct contribution from cellular metabolism

The third approach which has been used is to eliminate cellular metabolism and see whether for given ion gradients the cells still concentrate substrate to the same degree as normally. Potashner and Johnstone (16, 17) and Johnstone (19) have depleted Ehrlich ascites cells of ATP by incubating them with dinitrophenol and report that for comparable sodium and potassium gradients amino acid transport is less than in cells that have normal ATP levels. They have suggested that ATP might be needed to convert carrier to an activated form. If this is so, then ATP would still not make a direct energy contribution to the active transport but to the maintenance of the machinery of active transport. In other words, it would affect the coupling between the ion gradients and the amino acid movement. As such, this test does not distinguish between these two possibilities. It also suffers from much the same problems that plague the other tests. Determining true intracellular activities is still important but now the problem is also one of determining whether the intracellular activities are the same in differently treated cells; for example, the binding of ions may differ in ATP depleted and non-depleted cells. There is also the problem of excluding indirect effects resulting from the inhibition of the metabolic machinery, for example, changes in intracellular pH.

## Discussion

It is apparent, I think, that matters are still in an unsettled state in this field. Many of our problems would be settled if we could determine true cytoplasmic activities of the ions and amino acids. This is crucial for most of our tests. Even though all doubts would probably not be resolved, yet this one step would surely take us a long way in the resolution of these problems.

## Acknowledgements

The work reported here was supported in part by U. S. Public Health Service Research Grant No. CA-06734 from the National Cancer Institute.

I wish to thank Dr. J. Schafer for his critical review of the first draft of this paper.

## References

1. SCHULTZ, S. G. , CURRAN, P. F. : Coupled transport of sodium and organic solutes. Physiol. Revs. 50, 637-718 (1970).
2. EDDY, A. A. : A net gain of sodium ions and a net loss of potassium ions accompanying the uptake of glycine by mouse ascites-tumour cells in the presence of sodium cyanide. Biochem. J. 108, 195-206, (1968).
3. SCHAFER, J. A. , JACQUEZ, J. A. : Change in Na$^+$ uptake during amino acid transport. Biochim. Biophys. Acta 135, 1081-1083 (1967).
4. SCHAFER, J. A. : Thesis, University of Michigan, University Microfilms Ann Arbor, 1968.
5. GILLES-BAILLIEN, M. , SCHOFFENIELS, E. : Site of action of L-alanine and D-glucose on the potential difference across the intestine. Arch. Internat. de Physiol. Biochim. 73, 355-357 (1965).
6. WHITE, J. F. , ARMSTRONG, W. McD. : Membrane potentials in bullfrog small intestine: Effect of transported solutes. Biophys. Soc. Abstr. 36a, WPM-E9 (1970).
7. ROSE, R. C. , SCHULTZ, S. G. : Alanine and glucose effects on the intracellular electrical potential of rabbit ileum. Biochim. Biophys. Acta 211, 376-378 (1970).
8. ITOH, S. , SCHWARTZ, I. L. : Sodium and potassium distribution in isolated thymus nuclei. Am. J. Physiol. 188, 490-498 (1957).
9. ALLFREY, V. G. , MEUDT, R. , HOPKINS, J. W. , MIRSKY, A. E. : Sodium-dependent "transport" reactions in the cell nucleus and their role in protein and nucleic acid synthesis. Proc. Natl. Ac. Sci. (U. S. ) 47, 907-932 (1961).
10. McLAUGHLIN, S. G. A. , HURKE, J. A. M. : Sodium and water binding in single strated muscle fibers of the giant barnacle. Can. J. Physiol. Pharmacol 44, 837-848 (1966).
11. COPE, F. W. : NMR evidence for complexing of Na in muscle, kidney, and brain and by actomyosin. The relation of cellular complexing of Na to water structure and to transport kinetics. J. Gen. Physiol 50, 1353-1375 (1967).
12. ZADUNAISKY, J. A. , GENNARO, J. F. jr. , BASHIRELAKI, N. , HILTON, M. : Intracellular redistribution of sodium and calcium during stimulation of sodium transport in epithelial cells. J. Gen. Physiol. 51, 290s-302s (1968).
13. PIETRZYK, C. , HEINZ, E. : (These proceedings).
14. EDDY, A. A. : The effects of varying the cellular and extracellular concentrations of sodium and potassium ions on the uptake of glycine by mouse ascites-tumour cells in the presence and absence of sodium cyanide. Biochem. J. 108, 489-498 (1968).

15. JACQUEZ, J. A. , SCHAFER, J. A. : $Na^+$ and $K^+$ electrochemical potential gradients and the transport of $\alpha$ -aminoisobutyric acid in Ehrlich ascites tumor cells. Biochim. Biophys. Acta <u>193</u>, 368-383, (1969).
16. POTASHNER, S. , JOHNSTONE, R. M. : Cations, transport and exchange diffusion of methionine in Ehrlich ascites cells. Biochim. Biophys. Acta <u>203</u>, 445-456 (1970).
17. POTASHNER, S. J. , JOHNSTONE, R. M. : Cation gradients, ATP and amino acid accumulation in Ehrlich ascites cells. Biochim. Biophys. Acta <u>233</u>, 91-103 (1971).
18. SCHAFER, J. A. , HEINZ, E. : The effect of reversal of $Na^+$ and $K^+$ electrochemical potential gradients on the active transport of amino acids in Ehrlich ascites tumor cells. Biochim. Biophys. Acta <u>249</u>, 15 (1971).
19. JOHNSTONE, R. M. : (these proceedings).

## TABLE I

Comparison of maximum $c_i/c_e$, $Na_e/Na_i$ and $Na_e K_i/Na_i K_e$ in the steady state

| $Na_e$ | $Na_i$ | $K_e$ | $K_i$ | $c_e$ | $\dfrac{c_i}{c_e}$ | $\dfrac{Na_e}{Na_i}$ | $\dfrac{Na_e K_i}{Na_i K_e}$ |
|---|---|---|---|---|---|---|---|
| | | | | mmoles/kgm $H_2O$ | | | |
| 123.5 | 10.4 | 22.8 | 151.5 | 0.27 | 35.0 | 11.9 | 78.9 |
| 152.7 | 31.0 | 2.9 | 151.5 | 0.16 | 28.9 | 4.9 | 257.3 |
| 40.0 | 8.0 | 23.8 | 168.3 | 0.21 | 19.0 | 5.0 | 35.4 |
| 9.5 | 9.0 | 22.6 | 172.1 | 0.06 | 5.8 | 1.06 | 8.0 |
| 21.0 | 8.9 | 25.4 | 167.9 | 1.02 | 11.1 | 2.36 | 15.6 |
| 5.1 | 6.9 | 25.4 | 178.6 | 0.06 | 4.8 | 0.74 | 5.2 |
| 4.6 | 4.8 | 2.5 | 136.7 | 0.06 | 9.1 | 0.96 | 52.4 |

# Models of Coupling and their Kinetic Characteristics

E. Heinz
Institut für vegetative Physiologie der Universität Frankfurt/Main, Germany

The arguments used in the controversy about the interaction between non-electrolyte and electrolyte transport are largely based on kinetic observations. Any transport system in which such interaction occurs, e. g. by cotransport, is likely to display special kinetic features  which depend on the mechanism of coupling and on
the concentrations of the activating ions in the adjacent solutions; they are usually
interpreted in terms of preconceived models of hypothetical mechanisms. Most of
these models are based on the hypothesis of a "ternary complex" formed between
the transport carrier (X), the transported species (A), and the activating ion(s)
(e. g. $Na^+$). It should be kept in mind, however, that the formation of a ternary complex per se does not warrant the postulated transfer of energy, unless additional
assumptions are made, such as, for instance, that the formation of the complex
specifically modifies the carrier as to its affinities and/or mobility (1, 2). To provide a basis for the subsequent discussions,I shall try to give a brief survey of such
modifications that can be conceived, to scan them as to their presumable function
in the coupling process, and to derive the special kinetic features they presuppose
The latter appear to be sufficiently characterized by the following (standard) parameters:

a) the maximum velocity of flux ($J_{max}$),
b) the half-saturation constant ($K_m$) of this flux,
c) the maximum flux ratio ($f_{max}$), and
d) the stoichiometric ratio between the fluxes of substrate and co-substrate (r).

$f_{max}$ has to exceed unity if the model is to account for active accumulation of the
substrate, in that case excluding certain model types which permit mutual acceleration between substrate and co-substrate fluxes, but not active accumulation.
The stoichiometric ratio (r) is a very involved parameter whose intrinsic value,
owing to various uncontrolable circumstances, can hardly be measured precisely.
As a special session of this meeting has been scheluded for this topic, the discussion of r may be omitted here.

In order to keep the mathematical derivations simple,the following general assumptions are made:

1. The substrate (A) has the same concentration on either side of the membrane,
2. the modifier ($Na^+$) is present on the cis-side only,
3. trans-effects are absent.

Accordingly the parameters $J_a^{max}$ and $K_m$ refer to the unidirectional flux of A.

In the equations the following symbols are used:

$x_T$ = total concentration of carrier in the transport region;

$P_o^X$, $P_a^X$, $P_n^X$, $P_{an}^X$ are the velocity coefficients of the carrier X and its complexes
XA, $ANa^+$, $XNa^+A$, respectively, in the transport region; the superscript Y refers

to the corresponding coefficients of the Y carrier and its complexes, whenever they occur.

$K_a$, $K_n$ are the dissociation constants of XA and $XNa^+$, respectively, and $K'_a$, $K'_n$ are the corresponding dissociation constants of the ternary complex $(XNa^+A)$.

$$\varrho = \frac{K_a}{K'_a} = \frac{K_n}{K'_n}$$

Let us first consider co-transport only. According as the formation of the ternary complex modifies the affinity or the velocity of the carrier (X) we may distinguish between <u>affinity type models</u> and <u>velocity type models</u>:

I. <u>The affinity type models</u>: Here the binding of the co-substrate (e. g. $Na^+$) increases the affinity of the latter for the substrate (A), and <u>vice versa</u>. This effect <u>per se</u> suffices to account for the active accumulation of substrate via co-transport, even if the mobility of the carrier complex is unaffected by either binding. The affinity effect can be thought of as coming about in two ways, i. e. <u>directly</u> or <u>indirectly</u>. In the first case there is a real change in the affinities of the carrier as described. A typical example is the model proposed by Crane for the cotransport between sugars and Na ions in the intestine (3). In the second case, which we could also call <u>quasi-allosteric</u>, we assume two conformational states (X, Y) of the carrier to be equilibrium with each other. If only one of these (X) has a significant affinity for substrate and co-substrate, binding of either ligands will shift the equilibrium towards this conformational state. The kinetic behaviour is rather similar with both affinity type models. In either case we expect that only $K_m$ is affected by the coupling, not $J_{max}$. $f_{max}$ is clearly greater than 1, so that active accumulation can occur. The equation derived for the affinity type models under the simplifying assumptions mentioned above are the following:

a) direct effect: (Fig. 1, Ia)
  Special assumptions: 1. The velocity coefficient for all carrier complexes is equal to that of unloaded carrier, i. e. $P_a = P_n = P_{an} = P_0$. 2. $K_a > K'_a$, i. e. $\varrho > 1$

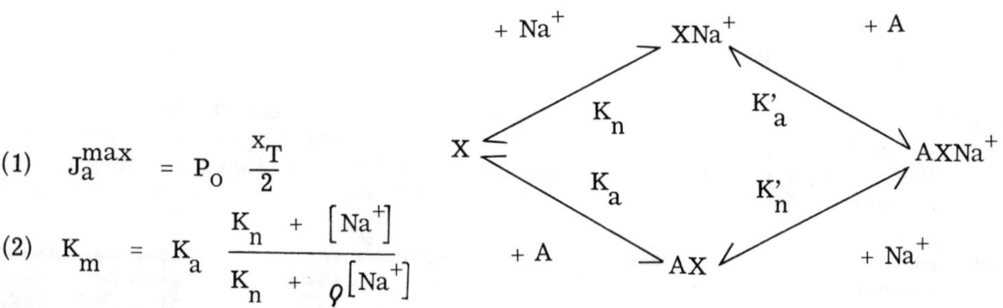


(1) $\quad J_a^{max} = P_0 \dfrac{x_T}{2}$

(2) $\quad K_m = K_a \dfrac{K_n + \left[ Na^+ \right]}{K_n + \varrho \left[ Na^+ \right]}$

These equations are essentially the same as those presented by Alvarez et al. for alanine transport in the intestine (5).

(3) $\quad f_{max} = \dfrac{K_n + \varrho \left[ Na^+ \right]}{K_n + \left[ Na^+ \right]}$

b) indirect (quasi-allosteric) effect (Fig. 1, Ib).
  Special assumptions: 1. The empty carrier species = $X_0$ and $Y_0$ are in equilibrium, i. e.
  $\dfrac{Y_0}{X_0} = L$; 2. only X binds $Na^+$ and A; $\varrho$ may be 1; 3. The

velocity coefficients of all carrier species are equal, i. e. $P_o^X = P_a^X = P_n^X = P_{an}^X = P_o^y$.

$$J_a^{max} = P_o \frac{x_T}{2} \tag{4}$$

$$K_m = K_a \frac{K_n (1 + L) + [Na^+]}{K_n + [Na^+]} \tag{5}$$

$$\int max = \frac{K_n (1 + L) + [Na^+] (1 + L)}{K_n (1 + L) + [Na^+]} \tag{6}$$

II. The velocity type models: Here the ternary complex has a higher velocity than the two binary ones between carrier and substrate or $Na^+$, respectively. This model in its pure form, i. e. with the affinity of the carrier for its ligands unaffected, can account for active accumulation only on the condition that

$$P_o \; x \; P_{ab} > P_a \; x \; P_b.$$

Otherwise $\int max$ will not exceed unity, as is shown below by equation 9a. Furthermore, the acceleration of the flux will not be greater than by the factor 2 unless

$$P_o > P_b.$$

In this model we expect that co-transport mainly affects $J_{max}$, with only a small, probably undetectable change of $K_m$. Various examples of such kinetic behaviour are described in the literature (for review see 1 and 4). Also the velocity effect may, in analogy to the affinity effect, be brought about directly and indirectly. In the indirect or quasi-allosteric case we again assume two conformational states of the carrier, but here they differ in their mobility. This model, however, in its pure form can be dismissed because it does under no circumstance account for accumulation ($\varrho_{max} = 1$), even though $J_{max}$ is increased by the coupling. Hence this coupling is merely catalytic, not energetic.

The standard parameters derived for the two velocity type models are as follows:
a) direct effect. (Fig. 1, IIa).
 Special assumptions: 1. $K_a = K_a'$ and $K_n = K_n'$, i. e. $\varrho = 1$; 2. The velocity coefficients of the different species of the carrier X are not equal, i. e. $P_o \neq P_a \neq P_n \neq P_{an}$.

$$J_a^{max} = P_o x_T \frac{P_a + P_{na} [Na^+] / K_n}{P_o + P_a + (P_o + P_n) [Na^+] / K_n} \tag{7}$$

$$K_m = K_a \frac{2 P_o + (P_o + P_n) [Na^+] / K_n}{P_o + P_a + (P_o + P_{na}) [Na^+] / K_n} \tag{8}$$

$$\int max = \frac{P_o (P_a + P_{na} [Na^+] / K_n)}{P_o P_a + P_a P_n [Na^+] / K_n} \tag{9}$$

$$\int_{max} > 1 \quad \text{if} \quad \frac{P_o \times P_{na}}{P_a \times P_n} > 1 \tag{9a}$$

b) indirect (quasi-allosteric) effect (Fig. 1, IIb).

Special assumptions: two conformations of carrier, X and Y, of which both have an equal affinity for A, whereas only X has an affinity for $Na^+$. The empty forms $X_o$ and $Y_o$ are in equilibrium, i.e. $\frac{Y_o}{X_o} = L$; The velocity coefficients of all species of X are equal ($P_X$), and so are those of Y ($P_y$), but $P_y < P_x$, i.e. $\frac{P_y}{P_x} = \varphi$, $\varphi < 1.$

$$J_a^{max} = \frac{X_T \, P_X \, (1 + \varphi L + [Na^+]/K_n)}{2(1 + L) + (1 + \frac{1 + L}{1 + \varphi L}) \, [Na^+]/K_n} \tag{10}$$

$$K_m = K_a \tag{11}$$

$$\int_{max} = 1 \tag{12}.$$

A counter-flow effect of a second substrate (K ion) can be easily implemented in any of the above models, e.g. by assuming that binding of this second cosubstrate

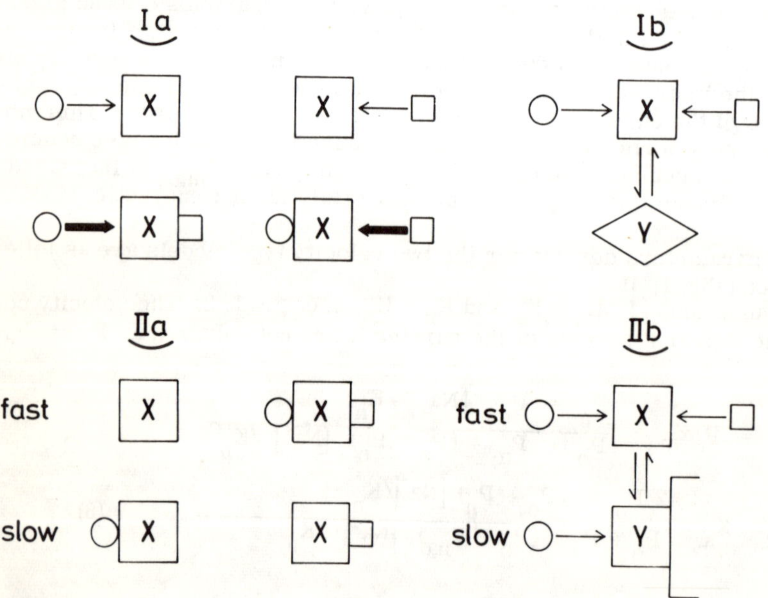


Fig. 1. Visualization of model types for Co-transport. X and Y are conformational states of carrier. Circles are the substrate (A), squares are the co-substrate ($Na^+$).

will have the opposite effect to that of the first.

There are many systems described in the literature in which the co-substrate affects both $J_{max}$ and $K_m$ markedly (6, 7). These cannot be satisfactorily accounted for by either of the above "pure" models. "Mixed type models" which combine affinity and velocity effects are therefore usually postulated. A more extended kinetical treatment of these models is given elsewhere (2).

## References


1. HEINZ, E.: in "Metabolic Pathways" Vol. VI (ed. L. Hokin) Academic Press New York 1972 (in press).
2. HEINZ, E., GECK, P., WILBRANDT, W.: Biochim. Biophys. Acta 255, 442 (1972).
3. CRANE, R. K.: Federation Proc. 24, 1000 (1965).
4. ALVAREZ, O., GOLDNER, A. M., CURRAN, P. F.: Amer. J. Physiol. 217, 946 (1969).
5. SCHULTZ, S. G., CURRAN, P. F.: Physiol. Rev. 50, 637 (1970).
6. INUI, Y., CHRISTENSEN, H. N.: J. Gen. Physiol. 50, 203 (1966).
7. EDDY, A. A.: Biochem. J. 108, 489 (1969).

# Lipid-Protein Interaction in Presence of Alkali-Cations

L. Bolis and C. Botré
Istituto di Fisiologia Generale, Università di Roma, Italy
Istituto Chimico-Farmaceutico, Università di Roma, Italy

An increasing interest has been devoted recently to the study of the interaction between lipids and proteins on physico-chemical grounds (Chapman, 1968; 1968a; 1969). The analysis of the macromolecular constituents of a biological membrane gives a picture of the lipid-protein interaction in an aqueous medium in the presence of different ions.

In addition, several ions are particularly involved in the functional activity of certain membranes, and the ionic composition is largely responsible for the non-equilibrium state of the living membrane.

The membrane shows the ability to select between very similar ions  and also to regulate the specific uptake of ions in different physiological conditions; nerve membrane in resting conditions is 30 times more permeable to $K^+$ than to $Na^+$, whereas during activity the nerve membrane is 10 times more permeable to $Na^+$ and $K^+$. Of course, these ions are very similar in their physical and chemical properties, as are other alkali ions ($Li^+$, $Rb^+$, $Cs^+$).

The ions have as rigid non-polarizable, monopolar spheres with differences in crystallographic radius ($Li^+$ O. 60 Å; $Na^+$ 0. 95 Å; $K^+$ 1. 33 Å; Rb 1. 48 Å; Cs 1. 69 Å) (Pauling, 1948).

In non-living systems such as minerals, ion-exchange resins and glass electrodes, an alkali cation selectivity with 11 permutation ways was ascertained (Diamond and Wright, 1968; 1969; Wright and Diamond, 1968). The interesting fact is that the selectivity observed in most biological systems corresponds to these 11 ways found for non-living systems. This situation led us to consider that the physical basis for this discrimination is similar for non-living and living systems. These five alkali cations can be permuted in 120 different ways.

The alkali cation selectivity equilibrium, which also predicts the transition sequence, was studied largely by Eisenman (1962, 1963, 1965, 1965a). The selectivity was ascribed to different attractive forces, mainly Coulombic forces, which are exerted by water on different cations, and by negative charges of the membrane. Anyway, the non-Coulombic forces involved could explain occasional deviations from patterns considered.

The interaction between protein and lipid in aqueous systems was studied on electrochemical aspect, in presence of different monovalent cations.


Experimental

The membranes used in this work were made out of polystyrene sulfonate embedded in a collodium matrix. The preparation and the use of such membranes as electrodes are described elsewhere (Clarke, 1954; Gregor and Sollner, 1954).

The polystyrene sulfonic acid had a molecular weight of about 90,000 and was prepared according to Neihof (1954). After purification the acid content of the polystyrene sulfonic acid was    4.80 meq/g, compared with a theoretical value of 5.43 meq/g, calculated for a linear polymer having one sulfonic acid group on each benzene ring.

Membranes were prepared by casting solutions of 4% collodium from a 1:2 mixture of alcohol and ether at various polystyrene sulfonic acid/collodium ratios.

Two types of membranes, $M_1$ and $M_2$, with different charge densities were prepared. The concentration of polystyrene sulfonic acid in membrane $M_1$ was $5 \times 10^{-4}$ eq/kg, and in the membrane $M_2$ $5 \times 10^{-1}$ eq/kg. Total exchange of hydrogen ions in the membranes with potassium ions was achieved by equilibrating the membranes in an approximate 1N potassium solution.

The coupled membrane system employed in potentiometric measurements may be schematized as follows:

Standard Calomel Electrode/KCl ref. sol./$M_1$/Soln in examen/$M_2$/KCl ref. sol/ Standard Calomel Electrode

where the potassium chloride reference solution was about 0.01N.

It has been shown theoretically that such an arrangement of membrane electrodes gives emf values which may be calculated by applying the membrane potential theory according to Teorell (1953) and Mayer and Sievers (1936).

However, by replacing the protein solution in the central compartment of the cell with potassium chloride solutions of different concentrations and by plotting the emf recorded at each concentration versus the logarithm of the mean ionic activity $(a_+)$ of potassium chloride in solution, a straight line was obtained (Fig. 1). The temperature was 22°C. The slope of the line in Fig. 1, i.e. 59 mV, originates from the slopes of the two membranes, i.e. 4 mV, for $M_1$ and 55 mV for $M_2$. Our aim was to obtain a system with a reproducible asymmetry.

In one set of experiments sonicated suspension of lecithin from egg (chromatographically pure, Gen. Biochem.) in alkali chloride $1 \times 10^{-4}$M) was used (different univalent cations for different experiments).

The central cell compartment was filled with saline solution of the same molarity as that containing L-$\alpha$-lecithin solution (1 mg/ml).

In other experiments we used the lipid part from a high density lipoprotein from human serum (HDL) (Scanu, 1966, 1967, 1968, 1970; Scanu and Granada, 1966; Bolis et al., 1967); the sonicated solution was prepared as indicated above and the same experimental procedure was used.

Crystallized bovine serum albumin obtained from BHD was dialyzed against water by allowing the solution to stand for five days at 4°C and lyophilized; the solution 0.5 mg/ml in the same alkali chloride, $1 \times 10^{-4}$ M, was used.

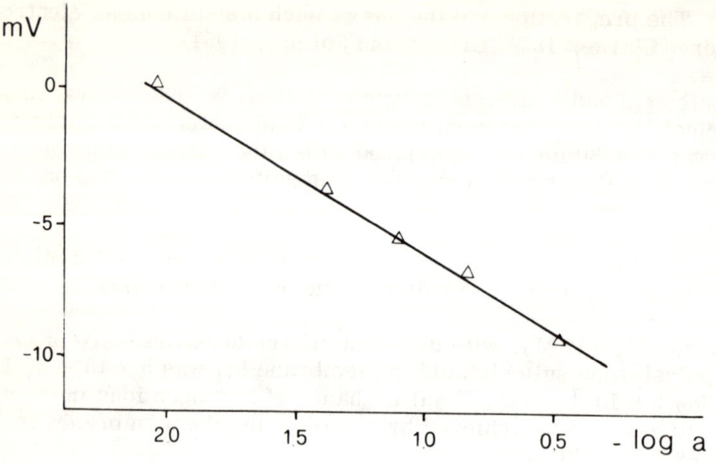

Fig. 1. emf (mV) is a function of the logarithm of the mean ionic activity a $\pm$ of a KCl solution placed in the central compartment of the cell and stepwise diluted. The temperature was 22°C

The proteic part delipidated from HDL in solution is 0.1 mg/ml in presence of the alkali as above.

The lipid of the proteic part was titrated against $\alpha$-lecithin and the emf derived was recorded.

## Results and Discussion

The stepwise addition, in the central compartment, of L-$\alpha$-lecithin (1 mg/ml) to solution of L-$\alpha$-lecithin, shows change in ionic activity (Fig. 2). It must be noted

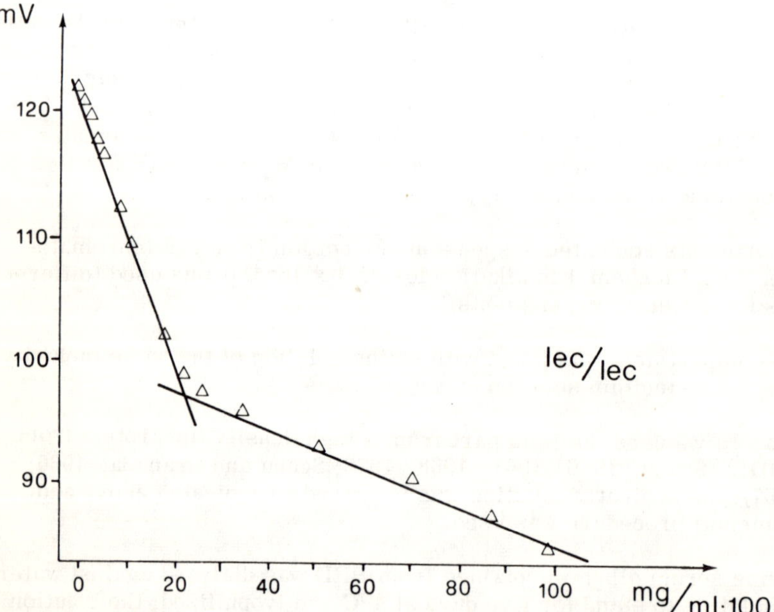


Fig. 2. Titration of lecithin with lecithin in $1 \times 10^{-4}$M KCl

that pure lecithin, although isoelectric over a wide pH range (Bangham et al.,
1958; Anderson and Pethica, 1956) swells in electrolyte solution to form a liquid
crystalline solution and "captures" some 0.1 $\mu$ mole of cation per $\mu$ mole of lecithin.

Recently, Bangham et al. (1965) demonstrated that there is a direct correlation
between the amount of alkali cation (K+) remaining after the dialysis in association
with the phospholipid structure: the amount is finite for pure lecithin, and increases
directly as a function of the surface charge increments either positive or negative.
The amount of cation is proportional to the physical properties of the lipid structure.

Change in behavior of the electromotive force when L-$\alpha$-lecithin was added in a
well stabilized suspension of phospholipids inducing a dilution in the system, could
be attributed to the release of trapped cations. Bangham et al. (1965 a) have shown
that both positively and negatively charged membranes release cations after transfer in an hypotonic medium.

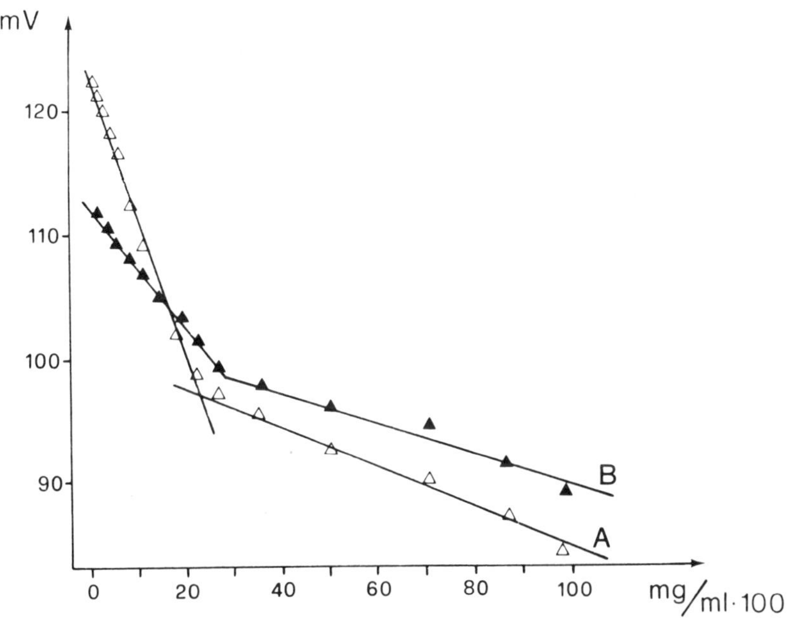


Fig. 3. Plot A (lec/lec) shows the electromotive force (in mV) recorded in an alkali chloride solution (1 x $10^{-4}$M LiCl) when increasing amounts of a concentrated
suspension (1 mg/ml) of L-$\alpha$-lecithin (in 1 x $10^{-4}$M LiCl) are added. Plot B
(BSA/lec) shows the electromotive force recorded in a bovine serum albumin
solution (0.5 mg/ml) in alkali chloride (1 x $10^{-4}$M LiCl) when increasing amounts
of L-$\alpha$-lecithin (in 1 x $10^{-4}$M LiCl) are added. The temperature was 25ºC $\pm$ 0.2,
pH 6.2 and constant.

No difference is noted in L-$\alpha$-lecithin solution against L-$\alpha$-lecithin solution when
the species of alkali cation changed in the medium.

The second part of the experiments deals with the titration of protein solution
against L-$\alpha$-lecithin.

First is the titration against bovine serum albumine (BSA) (Fig. 3).

Any further addition of lecithin after a certain value does not contribute to the electromotive force.

In agreement with the polyelectrolyte nature of the BSA it might be expected that some changes should occur in the binding of cations in solution. So the plots of electromotive force (in millivolts) recorded in a solution of BSA (0.5 mg/ml) added with L-$\alpha$-lecithin, show significant shifts which are in relation to the nature of the counterion present.

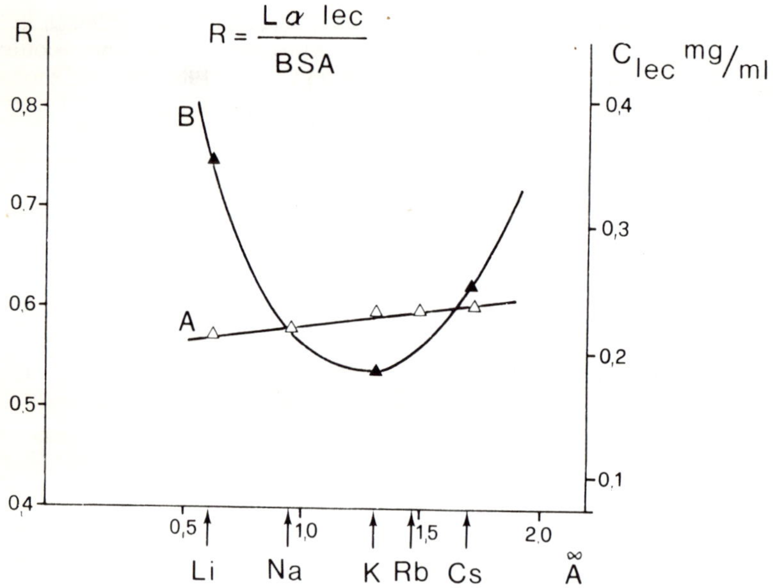


Fig. 4. Plot A, right ordinate: values of the L-$\alpha$-lecithin concentration at the breaking point of the plots obtained in different alkali chloride media, as a function of the crystallographic radius of the cations. Plot B, left ordinate: values of the ratio by weight of protein to lecithin at the breaking point of the plots obtained in different alkali chloride media, as a function of the crystallographic radius of the cations

During the experiments the concentration of the alkali chloride in the two external half cells was kept constant and equal to $1 \times 10^{-2}$M.

The titration of protein solutions with L-$\alpha$-lecithin is followed in presence of different cations like Li, Na, K, Rb, Cs, the resulting plot (Fig. 4) shows much the same shape, but the break point is shifted towards different values of the L-$\alpha$-lecithin concentration.

The break points expressed as a function of crystallographic radius of the cations show that in the case of BSA and lecithin the counterion binding does not follow the sequence to be expected merely on electrostatic grounds, and the binding of alkali cation exhibits a maximum for potassium ions.

The titration of apo-HDL from Scanu is also interesting. This protein part was

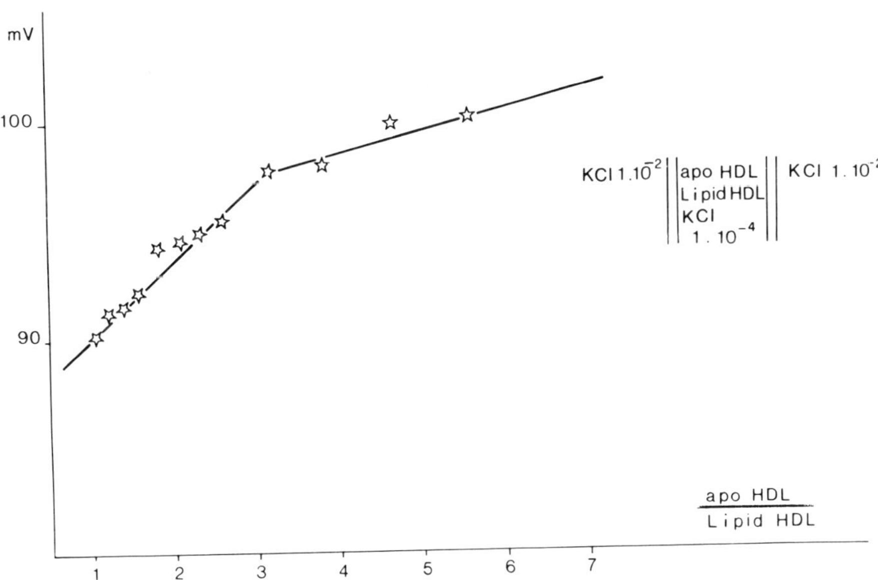



Fig. 5 A.  Titration of apo-HDL with lecithin (in KCl 1 x $10^{-4}$) crist.

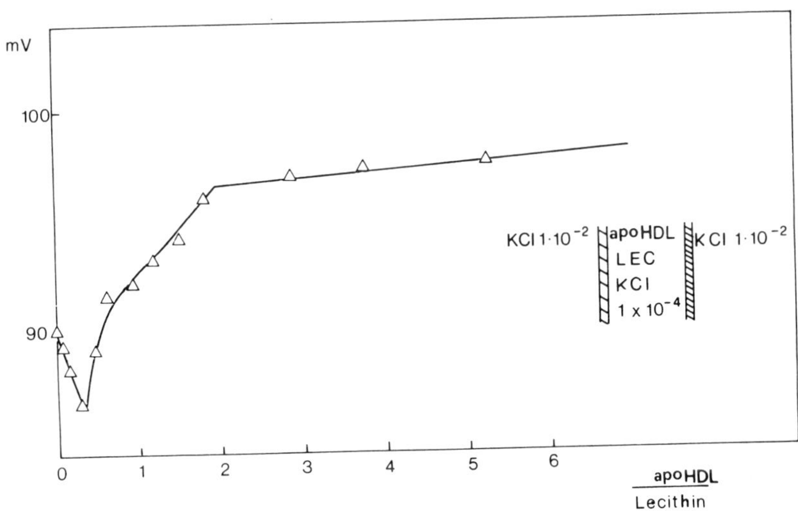



Fig. 5 B.  Titration of apo-HDL with lipid part from HDL

seen to react with this method in a ratio 2:1 both   with L-$\alpha$-lecithin and the lipid part extracted from the whole human lipoprotein in KCl (Fig. 5 A and B).  In this case the maximum of bindings is again for potassium, but the bindings for lithium and sodium are very similar (Fig. 6).

The same data are also obtained for apo-HDL and L-$\alpha$-lecithin.  Previously Scanu

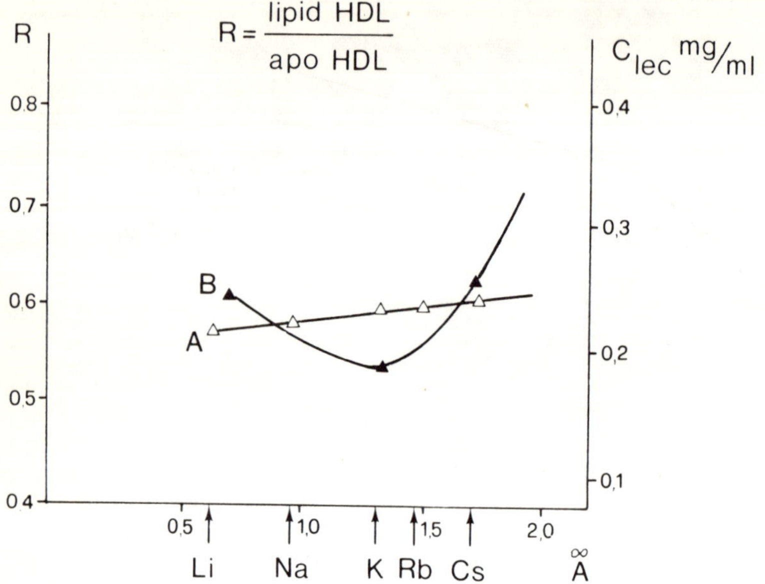



Fig. 6. Plot A, right ordinate: values of lipid part from HDL concentration at the breaking point of the plots obtained in different alkali chloride media, as a function of the crystallographic radius of the cations.
Plot B, left ordinate: values of the ratio by weight of protein to lecithin at the breaking point of the plots obtained in different alkali chloride media as a function of crystallographic radius of the cations

(1967) had pointed out that the apo-HDL binds lecithin similarly to the phospholipid part isolated from the lipoprotein. This means that this protein part shows a different behaviour in different ionic media during the titration with L-$\alpha$-lecithin.

The influence of counterion radius has been previously shown in other biological systems like polymerization of actin or BSA-histamine interaction (Botré et al., 1967, 1968).


## References


ANDERSON, P. J., PETHICA, B. A.: Proc. 2nd Int. Conf. Biochem. "Problems of Lipids", Butterworths, London (1956), p. 24.
BANGHAM, A. D., PETHICA, B. A., SEAMAN, G. V. F.: Biochem. J. 69, 12 (1958).
BANGHAM, A. D., STANDISH, M. M., WATKINS, J. C.: J. Mol. Biol. 13, 238 (1965).
BANGHAM, A. D., STANDISH, M. M., WEISSMAN, G.: J. Mol. Biol. 13, 253 (1965a).
BOLIS, L., BOTRÉ, C, BORGHI, S., MARCHETTI, M., in "Membrane Models and the Formation of Biological Membranes", p. 88. L. Bolis and B. A. Pethica (eds.), North Holland Publ. Co., Amsterdam 1968.
BOTRÉ, C., BORGHI, S., MARCHETTI, M.: Biopolymers 5, 483 (1967).
BOTRÉ, C., MARCHETTI, M., BORGHI, S.: Biochim. Biophys. Acta 154, 360 (1968).
CHAPMAN, D.: "Biological Membranes - Physical Fact and Function", Academic Press, New York - London 1968.

CHAPMAN, D.: in "Membrane Models and the Formation of Biological Membranes", p. 6, L. Bolis and B.A. Pethica (eds.), North Holland Publ. Co., Amsterdam (1968a).
CHAPMAN, D.: in "Structural and Functional Aspects of Lipoprotein in Living Systems", p. 3, E. Tria and A. Scanu (Eds.), Academic Press, New York - London (1969).
CLARKE, H.T.: "Ion Transport across Membranes", Academic Press, New York - London (1954).
DIAMOND, J.M., WRIGHT, E.M.: Fed. Proc. $\underline{27}$, 748 (1968).
DIAMOND, J.M., WRIGHT, E.M.: Ann. Rev. Physiol. $\underline{31}$, 581 (1969).
EISENMAN, G.: Biophys. J. $\underline{2}$, part 2, 259 (1962).
EISENMAN, G.: Bol. Inst. Estud. Med. Biol. $\underline{21}$, 155 (1963).
EISENMAN, G.: Proc. 23rd Intern. Congr. Physiol. Sci., Tokyo, Excerpta Med. Found., Amsterdam (1965).
EISENMAN, G.: Adv. Anal. Chem. Instr. $\underline{4}$, 213 (1965a).
GREGOR, H.P., SOLLNER, K.: J. Phys. Chem. $\underline{58}$, 409 (1954).
MAYER, K.H., SIEVERS, J.F.: Helv. Chim. Acta $\underline{19}$, 649 (1936).
NEIHOF, R.: J. Phys. Chem. $\underline{58}$, 916 (1954).
PAULING, L.: "The Nature of the Chemical Bond", Cornell University Press, Ithaca, New York, p. 450, 2nd Ed. (1948).
SCANU, A.M.: J. LipidRes. $\underline{7}$, 295 (1966).
SCANU, A.M., GRANADA, J.L.: Biochemistry $\underline{5}$, 446 (1966).
SCANU, A.M.: J. Biol. Chem. $\underline{4}$, 711 (1967).
SCANU, A.M.: in "Membrane Models and the Formation of Biological Membranes". L. Bolis and B.A. Pethica (eds.), p. 76, North Holland Publ. Co., Amsterdam (1968).
SCANU, A.M.: in "Permeability and Function of Biological Membranes", p. 239. L. Bolis et al. (eds.), North Holland Publ. Co., Amsterdam (1970).
TEORELL, T.: Progr. Biophys. Chem. $\underline{3}$, 305 (1953).
WRIGHT, E.M., DIAMOND, J.M.: Biochim. Biophys. Acta $\underline{163}$, 57 (1968).

# Neutral Amino Acids and the Ion Gradient Hypothesis

A. A. Eddy
Department of Biochemistry, University of Manchester Institute of Science
and Technology, Manchester, Great Britain

## Introduction

Current hypotheses about the mechanism of transduction of metabolic energy into
osmotic work during the transport of amino acids in mammalian tissues centre
about the question how sodium ions and potassium ions affect the process. The work
of Christensen and his colleagues early indicated that, when mouse ascites tumour
cells absorbed various neutral amino acids, the cellular content of $Na^+$ increased
and that of $K^+$ decreased (reviewed by Christensen, 1970). The time scale and stoichiometry of the ionic changes followed a complex pattern in relation to the amino
acid movements and it was not easy to interpret the results. Nevertheless Riggs,
Walker and Christensen (1958) were later able to show that the amount of glycine
accumulated by the mouse cells during one hour at $37^{\circ}C$ was progressively lowered
when cellular $K^+$ ions were replaced by $Na^+$ ions to an increasing extent. These and
other observations led to the now familiar hypothesis, which has been applied to
various mammalian tissues, that the spontaneous movements of $Na^+$ ions or $K^+$
ions across the cell membrane, down their respective cencentration gradients,
might drive the amino acid into the tumour cells without the direct intervention of
ATP. The concept is conveniently referred to as the ion gradient hypothesis. Though
he had no evidence implicating protons in the system, Christensen (1960) envisaged
the possibility that a concentration gradient of other ions such as $H^+$ might also
serve to propel the amino acid up its own concentration gradient. It is interesting
that recent evidence suggests that protons may be so involved in the transport of
amino acids in certain yeasts (Eddy and Nowacki, 1971).

A number of specific questions naturally arise in attempting to apply the ion gradient hypothesis to the mammalian systems. (1) Are both $Na^+$ and $K^+$ implicated
and how do they interact with the amino acid fluxes into and out of the cells? (2)
What is the effect on the influx and efflux kinetics of depriving the system of ATP?
(3) How does the steady state distribution of amino acid vary with the magnitudes
of the concentration gradients of $Na^+$ and $K^+$ between the cells and their environment? We shall consider these questions principally in relation to the mouse ascites
tumour cells, noting where appropriate any similarities to the behaviour of pigeon
erythrocytes and the behaviour of various intestinal preparations, the other systems
that have been extensively investigated. (reviewed by Schultz and Curran, 1970).

## The ions as cosubstrates

Riggs et al. (1958) pointed out, in their original statement of the gradient hypothesis, that the simplest mechanism for coupling the movement of the alkali cation to
that of the amino acid required that the carrier initially binding the amino acid molecule also associated with $Na^+$ or $K^+$. That $Na^+$ ions were of critical importance
in this connexion was shown by Kromphardt, Grobecker, Ring and Heinz (1963).

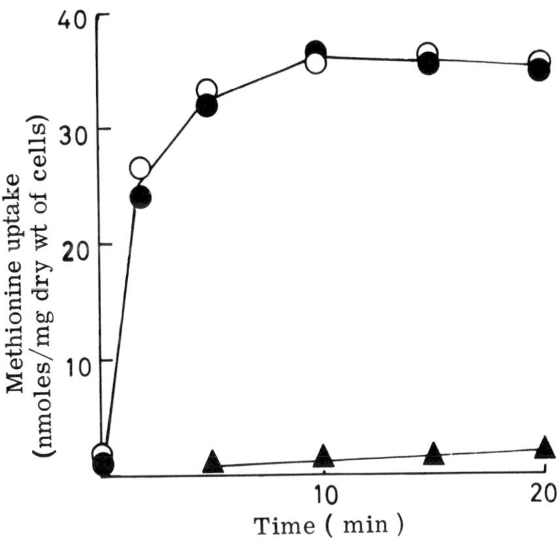

Fig. 1. Uptake of L-methionine from a 1 mM solution by the starved mouse tumour cells in the presence of 2 mM sodium cyanide. O, ninhydrin assay; ●, assay with $^{14}$C methionine; extracellular $Na^+$ was 160 m-equiv/1 in both cases. ▲ , assay with $^{14}$C methionine, $Na^+$ replaced by $K^+$. The endogenous amino acid pool was initially depleted by subjecting the tumour cells to a hypotonic shock, in a potassium Ringer of half the normal tonicity, for 10 min at 37$^{\circ}$C. They were then transferred to the standard potassium Ringer for 25 min. (Observations of M. Morville)

| | |
|---|---|
| ordinate | methionine uptake (nmole/mg dry wt of cells) |
| abscissa | time (min) |

They observed that lowering the concentration of $Na^+$ ions in the extracellular phase dramatically lowered the influx of glycine. In practice the absorption of glycine by the mouse cells from relatively dilute solutions ($< 2$ mM) virtually stops in the absence of $Na^+$. Similar observations have been made with 2-aminoisobutyric acid (Jacquez, Sherman and Terris, 1970). Conflicting reports about the magnitude of the effect of $Na^+$ on the behaviour of L-phenylalanine, L-leucine and certain other amino acids (see Schultz and Curran, 1970) with the mouse tumour cells have not been resolved. It is possible that some of these amino acids exchange with endogenous amino acids independently of the presence of $Na^+$ ions and to a variable extent in different cell preparations (Johnstone and Scholefield, 1965). Fig. 1 shows that the strain of mouse tumour cells used in the author' s laboratory absorbed L-methionine much faster when the Ringer contained $Na^+$ as opposed to $K^+$. Exchange with endogenous amino acids appeared not to be involved under the conditions described in the legend to Fig. 1.

Work in various laboratories has shown that the flow of amino acid stimulated by extracellular $Na^+$ is accompanied by an accelerated influx of $Na^+$ ions. For instance, Schafer and Jacquez (1967) found that about 1 extra equivalent of $^{22}Na^+$ accompanied each equivalent of 2-aminoisobutyric acid into the tumour cells. The question then arises whether a net uptake of $Na^+$ occurs with the amino acid, or whether both the influx and the efflux of $Na^+$ are faster. The issue was resolved by using flame photometry to show that when the tumour cells were depleted of ATP about 0.9 extra equivalent of $Na^+$ were absorbed with each equivalent of glycine absorbed from an 11 mM solution (Eddy, 1968a). About 0.6 extra equivalent of $K^+$, in excess of the number lost from the controls without glycine, were simultaneously ejected from the tumour cells. A net flow of $Na^+$ ions thus accompanies the amino acid into the tumour cells under these conditions and is partially neutralized by a net flow of $K^+$ ions in the opposite direction.

The above situation may be contrasted with the behaviour of pigeon erythrocytes. In that system both the influx and the efflux of $Na^+$ ions, as well as the efflux of endogenous amino acids were stimulated when L-alanine was absorbed during energy metabolism (Wheeler and Christensen, 1967). The uptake of $Na^+$ with glycine shows different characteristics (Vidaver, 1964). About two extra equivalents of $Na^+$ were absorbed during the net uptake of glycine in circumstances where exchange with the endogenous amino acids was probably not involved.

We may now return to the behaviour of the mouse cells with glycine. Table 1 summarizes the kinetic constants defining the binding both of $Na^+$ ions and of glycine itself to the hypothetical amino acid carrier (E). The estimates were obtained by studying the effect on the initial rate of uptake of the amino acid of varying the values of $[Na^+]$, $[K^+]$ and $[gly]$ in the extracellular phase. Table 1 shows, in agreement with the observed displacement of $K^+$ ions in the presence of glycine, that $K^+$ ions interact with the system in an interesting way. They appear both to compete with $Na^+$ for the site where that ion binds and to lower the affinity of the carrier complex for glycine. One assumption underlying these computations is that glycine crosses the cell membrane only as the ternary complex ENaGly. Another important one is that the various ligands bind to the carrier much more rapidly than the resultant complexes traverse the membrane. On that basis, the stoichiometry of the induced ionic movements, together with an analysis of the so-called trans effects of the cations on glycine influx, provides information about the mobilities of the various carrier species (Eddy, 1968a; Eddy and Hogg, 1969). It was concluded that ENa crosses the cell membrane relatively slowly, or not at all, and that both E itself and EK cross relatively rapidly.

A detailed analysis of the role of $Na^+$ in the transfer of L-alanine across the mucosal border of the intestinal epithelial cell of the rabbit ileum has led to a model differing in a number of ways from the above scheme (Schultz and Curran, 1970). In particular, the carrier appears to transfer alanine in two forms which, in the above terminology, may be represented by the complexes EAla and ENaAla. With the same notation the complex $ENa_2Gly$ has likewise been postulated for the pigeon erythrocytes. A further variation from the behaviour of the mouse cells was encountered, in that $K^+$ appeared not to inhibit the functions of $Na^+$ (Vidaver, 1964). The differences in stoichiometry exhibited by the various systems are probably not to be taken as evidence against the cosubstrate role of $Na^+$ in the transport of these amino acids, but rather as providing a range of models based on the same fundamental postulates.


## Application of carrier theory

The evidence defining the binding constants illustrated in Table 1, together with the evidence that the flow of glycine into the tumour cells induces an influx of $Na^+$ and an efflux of $K^+$, leads to the hypothesis of the carrier complex ENaGly. Bearing in mind that qualitatively similar interactions occur when the system is deprived of ATP, the simplest assumption to make is that the carrier binds the various ligands as readily on one side of the membrane as on the other. Some differences in the kinetic constants applying to the binding of $Na^+$ and glycine on the two sides of the cell membrane of the pigeon erythrocyte have in fact been described by Vidaver and Shepherd (1968). Whether the above assumption holds rigorously in the tumour system is not known. It seems to be a valid approximation however. The effect of $K^+$ might be exerted in either or both of two ways. (1) The presence of $K^+$ might increase the efficiency of the coupling between the $Na^+$ gradient and the amino acid gradient. (2) The energy inherent in the $K^+$ ion gradient itself might be coupled to the flow of the amino acid. The second possibility is the one that seems at present more likely to apply to the mouse tumour cells. The effect of varying both the cellular and extracellular concentrations of $Na^+$ and $K^+$ on the distribution of the amino acid can be predicted by extending the equations used in the classical model of facilitated diffusion (Wilbrandt and Rosenberg, 1961).

When E, ENa and EK behave as described above and the distribution of glycine between the cellular and the extracellular compartments, denoted respectively by the subscripts i and o, is determined solely by the prevailing concentration of $Na^+$ and $K^+$, we find that

$$\frac{[Gly]_i}{[Gly]_o} = \frac{[Na^+]_o}{[Na^+]_i} \cdot \frac{(1 + \theta [K^+]_i)}{(1 + \theta [K^+]_o)} \qquad (1)$$

The parameter $\theta$ is defined as the factor $k^K/k^c k_2$, where $k^K/k^c$ is the relative rate of transfer of EK as opposed to E and $k_2$ is the dissociation constant of the EK complex (Eddy, 1968a). Equation 1 neglects the effect of the membrane potential which we have attempted to take into account in our most recent work as follows. We introduce the factors P and P' to allow for the possibility that the unloaded carrier moves inwards across the cell membrane at the relative rate $P k^c$ and moves outwards at the rate $P' k^c$. The relation between the ion gradients and the glycine gradient then becomes

$$\frac{[Gly]_i}{[Gly]_o} = \frac{[Na^+]_o}{[Na^+]_i} \cdot \frac{P' + \theta [K^+]_i}{P + \theta [K^+]_o} \qquad (2)$$

A possible way of estimating P and P' is to regard E as an anion carrying a single charge and ENa as a neutral species. Both factors are then estimated by using the assumptions underlying the Goldman relationship (Katz, 1966) and the likely values of the membrane potential (cf. Britton, 1965). For instance, according to Simonsen and Nielsen (1971), the membrane potential corresponds to a chloride ratio ($[Cl^-]_o / [Cl^-]_i$) of about 2.6. We find that P' would then be about 1.55 and P would be about 0.60, the terms of the hypothesis requiring that the ratio P'/P is also 2.6.

An interesting aspect of the model leading to equation 2 is that when the electrical potential of the cellular phase becomes more negative, as it might do if an electrogenic pump were operating, the system would tend to become uncoupled from the potassium ion gradient. The amino acid pump might then appear to depend directly on the supply of metabolic energy to the electrogenic pump. Whether this possibility is relevant to the mouse-tumour system is not known but, if the proton pump in yeast is electrogenic, it may help to explain by Eddy and Nowacki (1971) only observed coupling of amino acid transport to $K^+$ in yeast when the cells were deprived of ATP.

Another complication arising in practical applications of either equation 1 or 2 is that little is known about the activity coefficients of the ions in the cellular phase, or about the distribution of $Na^+$ and $K^+$ between various cellular compartments. We have accordingly found it expedient (Reid and Eddy, 1971) to multiply the right hand side of equation 1 by a correction factor f ($1 < f < 2$). The applications of these numerical methods have still to be explored.

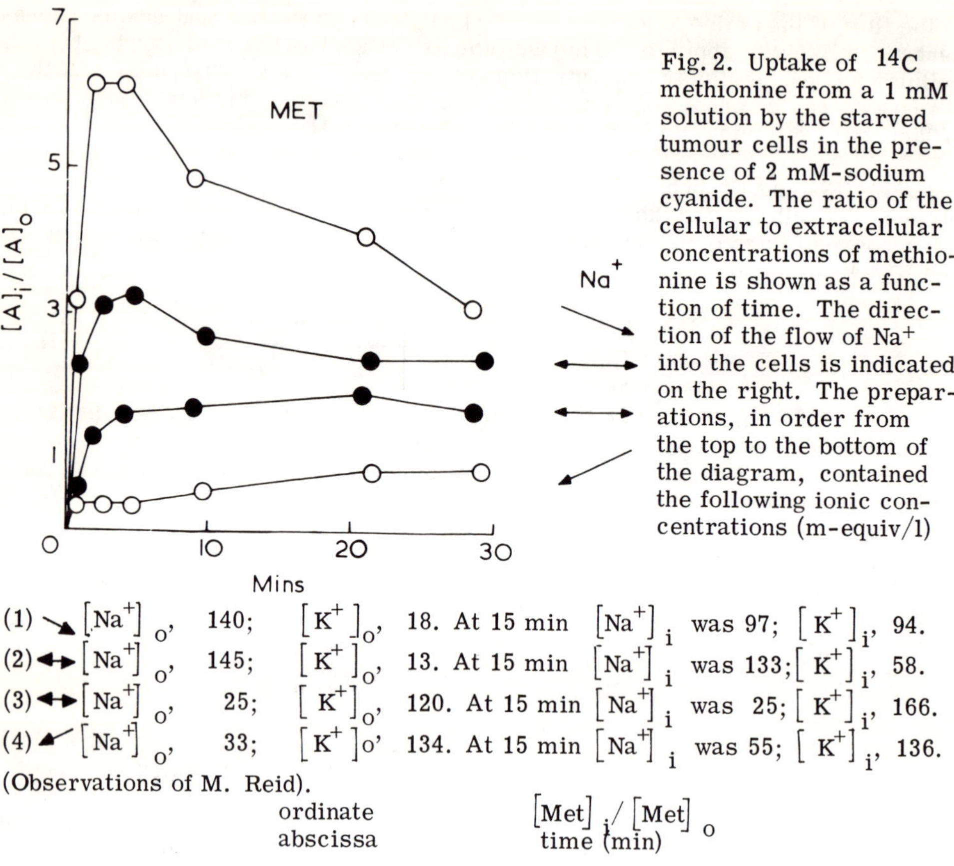



Fig. 2. Uptake of $^{14}C$ methionine from a 1 mM solution by the starved tumour cells in the presence of 2 mM-sodium cyanide. The ratio of the cellular to extracellular concentrations of methionine is shown as a function of time. The direction of the flow of $Na^+$ into the cells is indicated on the right. The preparations, in order from the top to the bottom of the diagram, contained the following ionic concentrations (m-equiv/l)

(1) $\left[Na^+\right]_o$, 140; $\left[K^+\right]_o$, 18. At 15 min $\left[Na^+\right]_i$ was 97; $\left[K^+\right]_i$, 94.

(2) $\left[Na^+\right]_o$, 145; $\left[K^+\right]_o$, 13. At 15 min $\left[Na^+\right]_i$ was 133; $\left[K^+\right]_i$, 58.

(3) $\left[Na^+\right]_o$, 25; $\left[K^+\right]_o$, 120. At 15 min $\left[Na^+\right]_i$ was 25; $\left[K^+\right]_i$, 166.

(4) $\left[Na^+\right]_o$, 33; $\left[K^+\right]_o$, 134. At 15 min $\left[Na^+\right]_i$ was 55; $\left[K^+\right]_i$, 136.

(Observations of M. Reid).

ordinate     $\left[Met\right]_i / \left[Met\right]_o$

abscissa     time (min)

## Experimental tests

Vidaver (1964), working with reversibly hemolyzed pigeon erythrocytes, obtained the first clear indication that, when ATP was lacking, the distribution of glycine between the cellular and extracellular phases varied systematically with the magnitude of the sodium gradient ( $\left[Na^+\right]_o / \left[Na^+\right]_i$ ). He found that when $\left[Na^+\right]_o >$ $\left[Na^+\right]_i$ then $\left[gly\right]_i > \left[gly\right]_o$. Correspondingly, reversing the sodium gradient ( $|Na^+|_o < |Na^+|_i$ ) reversed the direction of the glycine gradient. The exact relationship between these quantities has not been established in this system and is complicated by the tendency of the resealed erythrocytes to leak solutes.

Work on similar lines with the mouse tumour cells deprived of ATP, either in the presence of cyanide or cyanide plus deoxyglucose (Eddy, Mulcahy and Thomson, 1967; Eddy, 1968a, b; Eddy and Hogg, 1969), revealed a similar correlation between the direction and magnitude of the sodium gradient and the direction and magnitude of the glycine gradient. The application of equation 1 showed further that some degree of coupling to the potassium gradient was also involved.

The question naturally arises whether ATP formation had in fact stopped in these preparations. The relevant points are, first, that the cellular ATP content was relatively small and, second, that both respiration and glycolysis were largely inhi-

Fig. 3. The relative magnitudes of the methionine gradient and the sodium ion gradient during respiration (see text). $[Na^+]_o$ and $[Na^+]_i$ were varied as in the series of observations shown in Fig. 6 of Eddy (1968b). $[Met]_o$ was 1 mM. (Observations of M. Reid).

ordinate   log $[Met]_i / [Met]_o$

abscissa   log $[Na^+]_o / [Na^+]_i$

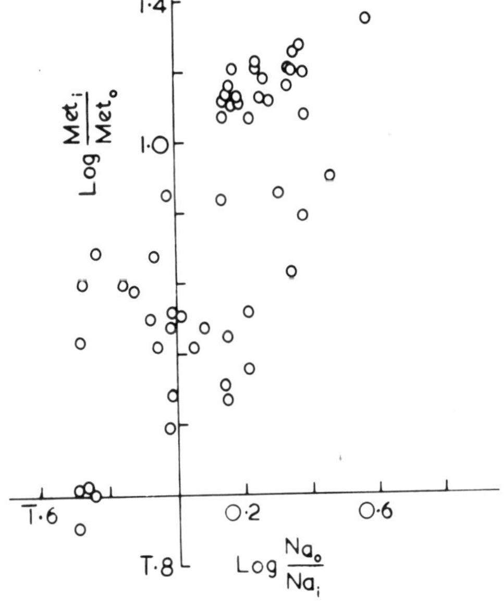


bited. It was more difficult to decide whether the accumulation of glycine in the presence of the metabolic inhibitors depended on an exchange reaction with the endogenous complement of amino acids. This possibility has now been made unlikely on the basis of observations like those shown in Fig. 1.

Whereas our earlier work was done with glycine we have since found that L-methionine behaves similarly. Fig. 2 shows some typical observations relating the effect of varying the initial ionic gradients between the cellular and extracellular phases to the progress of the absorption of methionine by preparations of the tumour cells depleted of ATP. While the considerable influence of $Na^+$ is at once apparent, the relative importance of $Na^+$ and $K^+$ can only be assessed by making a more detailed analysis, preferably with the help of the mathematical models. One way of proceeding is to compare the magnitudes of the sodium and methionine gradients in various situations where both have been deliberately varied. Fig. 3 illustrates a series of such observations, in this case made with respiring cells. The two parameters are expressed on a logarithmic scale for convenience. Clearly, they are only weakly correlated. Fig. 3 shows that a two fold sodium gradient was regularly associated with a twenty fold gradient of methionine concentration. Were the movement of $Na^+$ alone supposed to drive methionine into the tumour cells, at least 4 such $Na^+$ ions would be required to accompany each equivalent of methionine. One might alternatively suppose that the effective activity coefficient of the $Na^+$ ions in the cellular phase was much lower than it appeared to be, either because a membrane potential of about 60 mv was present, or because much of the cellular $Na^+$ was confined to a separate compartment. In general neither possibility seems very plausible as applied to the mouse tumour cells.

Observations like those shown in Fig. 3, obtained with cellular preparations containing varying amounts of ATP, have led Potashner and Johnstone (1971) to reject the notion that the sodium gradient plays a major role in determining the distribution in the Ehrlich strain of mouse tumour cells. These workers took little account of the possible role of the potassium gradient, however, attaching more

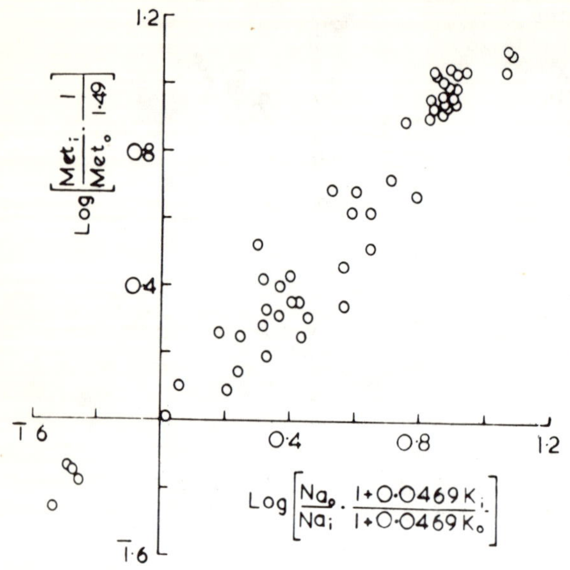



Fig. 4. The observations illustrated in Fig. 3 now represented in the form of equation 1, which takes into account both the sodium ion and the potassium ion gradient. Adapted from Reid and Eddy (1971).

ordinate: $\log [\text{Met}]_i / 1.49 [\text{Met}]_o$

abscissa:

$$\log \frac{[\text{Na}^+]_o}{[\text{Na}^+]_i} \cdot \frac{1 + 0.0469 [\text{K}^+]_i}{1 + 0.0469 [\text{K}^+]_o}$$

significance to the positive correlation they found between the cellular ATP content and the magnitude of the methionine gradient. They concluded that the hydrolysis of ATP itself was more likely to provide the primary force driving the amino acid pump than was the sodium gradient.

How does Potashner and Johnstone's hypothesis apply to Fig. 3? The way in which the cellular ATP content varied with the methionine gradient is not known, unfortunately, although it seems likely that all the preparations in fact contained relatively large amounts of ATP. We may instead ask how the methionine was distributed under the combined influence of both the sodium ion and the potassium ion gradients. Fig. 4 is concerned with that issue. It shows that the data represented in Fig. 3 roughly fit the mathematical model based on equation 1, but only provided the sodium gradient is multiplied by the relatively small correction factor of 1.49 (factor $\underline{f}$ of Reid and Eddy, 1971). The possible biological significance of $\underline{f}$ is discussed below. At this stage of the argument we can regard it as an empirical factor introduced so as better to reveal the extent to which the ionic gradients govern the accumulation of methionine. Meanwhile, we may note that the values of both $\Theta$ and the parameter $\underline{f}$ are chosen by a conventional least squares criterion. When $\theta$ is relatively large, close coupling with the potassium gradient is observed and the term $1 + \theta [\text{K}^+]_i / 1 + \theta [\text{K}^+]_o$ approaches the ratio $[\text{K}^+]_i / [\text{K}^+]_o$. The observed value of 0.047 for $\theta$ indicates that the methionine gradient was only partially coupled, however, to the potassium gradient in these circumstances. For example, $[\text{K}^+]_i / [\text{K}^+]_o$ was about 16 in one instance where a 6 fold, rather than a 16 fold, gradient of methionine concentration could be attributed to the potassium ion gradient.

By introducing the correction factor $\underline{f}$, we are thus led to the conclusion that the adjusted data conform roughly to equation 1. To that extent the system behaved as though the gradients of both $\text{Na}^+$ and $\text{K}^+$ were the main factors controlling the accumulation of methionine. In other words the analysis appears to make it unnecessary to suppose, as Potashner and Johnstone (1971) have done, that the methionine pump used ATP directly. On the other hand, the functioning of the sodium pump, that pre-

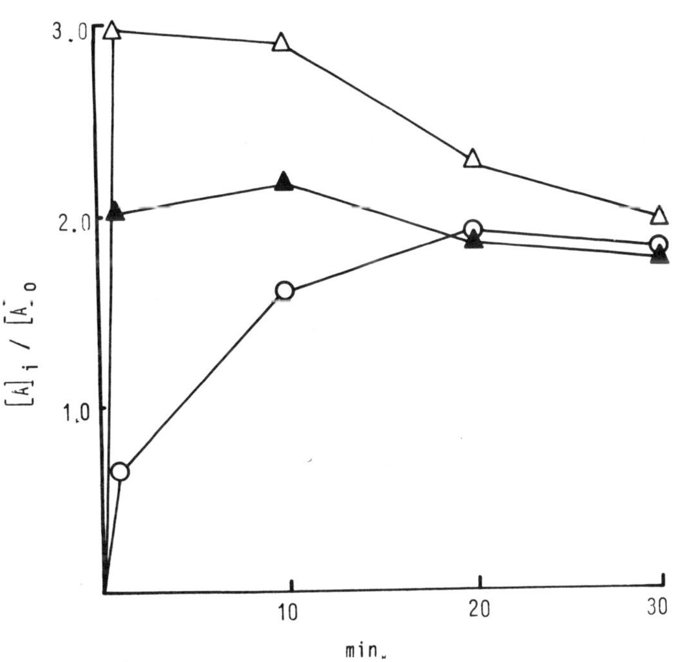



Fig. 5. The accumulation of glycine (▲), L-methionine (△) and L-lysine (○) when the sodium ion and potassium ion concentration gradients were small and the tumour cells were depleted of ATP. The extracellular phase contained a 1 mM solution of the test amino acid, labelled with $^{14}$C, 2 mM-sodium cyanide and 10 mM-2-deoxy-glucose (Eddy and Hogg, 1969). The ratio of the concentrations of $^{14}$C in the cellular and extracellular phases ( $[A]_i / [A]_o$) was determined at the times indicated. $[Na^+]$ was 116 m-equiv/l and $[K^+]$ was 34 m-equiv/l in the extracellular phase. At 30 min, cellular $[Na^+]$ in each of the three preparations was in the range 142-149 m-equiv/l and cellular $[K^+]$ was in the range 51-53 m-equiv/l

sumably maintained the ionic gradients in at least some of the circumstances in which the methionine gradient was measured, may be expected to have utilized ATP.


## The membrane potential and the significance of f

The validity of the above argument obviously depends on the significance attached to factor f. It was originally introduced because $[Met]_i / [Met]_o > 1$ when $[Na^+]_i$ and $[Na^+]_o$, on the one hand, and $[K^+]_i$ and $[K^+]_o$, on the other hand, were approximately equal (Reid and Eddy, 1971). The work of Potashner and Johnstone (1971) had first drawn our attention to this possibility. Some relevant observations made by L. Gibb, with preparations of the tumour cells depleted of ATP, are illustrated in Fig. 5. It will be seen that the respective concentrations of L-methionine, glycine and L-lysine that eventually accumulated in the tumour cells were almost twice those in the extracellular phase, even though the ionic concentration gradients of Na$^+$ and K$^+$ had largely dissipated. Such behaviour might be attributed to various factors, including the possibility that cellular Na$^+$ was held in discrete compartments. However, we favour the working hypothesis that the accumulation of the amino acids under these conditions was governed by a membrane potential of the Donnan type. An extension of that hypothesis leads to the possibility that the factor $\underline{f}$ may simply provide a rough method of correcting the predictions based on equation 1 for the effect of the membrane potential on the accumulation of the amino acid. The latter effect may indeed be more accurately described by equation 2, so that the multiplication by $\underline{f}$ is perhaps approximately equivalent to the introduction of P' and P. Further work is needed to clarify this important issue, which might obviously lead to a reassessment of the quantitative effects of $K^+$ in the system (cf. Eddy, 1968b).

## Regulation of the coupling mechanism

The conflict in the literature about the functions of ATP in amino acid transport makes it tempting to suppose that the validity of the gradient hypothesis might better be explored by examining the properties of the cells depleted of ATP. The sodium pump would then stop working and, in so far as this affected the magnitudes of $[Na^+]_i$ and $[K^+]_i$, one might attempt, by making use of both Fig. 4 and the appropriate mathematical equations, to predict how the methionine gradient might change in the presence of metabolic inhibitors that stopped energy metabolism. There are now indications that the real position is more complicated. One critical observation is that a concentration of cyanide, with or without the addition of deoxyglucose, that largely inhibits ATP formation lowers the accumulation of both glycine and methionine to a greater extent than would be predicted on the above basis (Reid and Eddy, 1971). We have suggested that the effect involves a partial loss of coupling of the amino acid pump to the potassium ion gradient (Table 2).

These observations have led us to the view that ATP, or some related compound, may stimulate the uptake of the amino acids by regulating the efficiency of the coupling mechanism controlling the flow of $K^+$ through the amino acid carrier. On that interpretation the passage of the amino acid through the carrier would not involve a simultaneous stoichiometrical hydrolysis of ATP. The suggested regulatory role of the hypothetical product of energy metabolism, which might indeed be ATP itself, might be analogous to the action of an allosteric ligand. Alternatively, by analogy with more familiar enzyme systems, it might depend on the phosphorylation of the carrier E, in a reaction not involving the transfer of the amino acid through the carrier system.

It will be apparent that, in principle, the coupling of the flow of $Na^+$ with that of the amino acid might also be regulated. One can envisage that the relative mobilities of the species E and ENa might vary with the physiological state of the cells. However, we have no experimental support for that possibility.

The existence of the above effects complicates the interpretation of the actions of metabolic inhibitors on the mouse tumour cells in terms of the ion gradient hypothesis. For instance, ouabain might inhibit the sodium pump and thereby affect not only the cellular ion content, but also the supply of ATP, or ADP, to the systems controlling the coupling between the potassium gradient and the amino acid gradient. The proposal by Kimmich (1970) that ouabain may directly inhibit amino acid transport in certain intestinal preparations seems relevant in this connexion. A further complication might arise if the metabolic control of the coupling depended on variables that were themselves influenced by the ionic composition of the cells. A rigorous analysis of the experimental data in terms of the mathematical models may help to resolve such matters.

TABLE I

Apparent dissociation constants of the carrier complex with $Na^+$, $K^+$ and glycine. Adapted from Eddy, Mulcahy and Thomson (1967).

| Ligand bound | | Dissociation constant (m-equiv/l) |
|---|---|---|
| $Na^+$ | | 37. 5 |
| $K^+$ | | 76 |
| Glycine | | 6. 4 |
| $Na^+$, then glycine | | 3. 5 |
| Glycine, then $Na^+$ | | 21 |
| $K^+$, then glycine | | 92 |
| Glycine, then $K^+$ | (about) | 1000 |

TABLE II

The effect of energy metabolism on the parameter $\theta$ expressing the magnitude of the coupling between amino acid transport and the potassium ion gradient (Reid and Eddy, 1971).

| | Respiration | $10^3\ \theta$ |
|---|---|---|
| 1 mM-glycine | + | 28 |
| | − | 11 |
| 1 mM-methionine | + | 47 |
| | − | 10 |

# References


BRITTON, H. G. : J. Theoret. Biol. $\underline{10}$, 28 (1965).
CHRISTENSEN, H. N. : Advan. Protein Chem. $\underline{15}$, 239 (1960).
CHRISTENSEN, H. N. : in "Membranes and Ion Transport", vol. 1 (ed. by Bittar) Wiley-Interscience, London 1970, p. 365.
EDDY, A.A. , MULCAHY, M. F. , THOMSON, P. J. : Biochem. J. $\underline{103}$, 863 (1967).
EDDY, A.A.: Biochem. J. $\underline{108}$, 195 (1968a).
EDDY, A.A.: Biochem. J. $\underline{108}$, 489 (1968b).
EDDY, A.A. , HOGG, M. C. : Biochem. J. $\underline{114}$, 807 (1969).
EDDY, A.A. , NOWACKI, J.A. : Biochem. J. $\underline{122}$, 701 (1971).
JACQUEZ, J.A. , SHERMAN, J. H. , TERRIS, J. : Biochim. Biophys. Acta $\underline{203}$, 150 (1970).
JOHNSTONE, R. M. , SCHOLEFIELD, P. G. : Biochim. Biophys. Acta $\underline{94}$, 130 (1965).
KATZ, B.: Nerve, Muscle and Synpase, McGraw-Hill Inc. , New York 1966, p. 60.
KIMMICH, G.A.: Biochemistry, Easton $\underline{9}$, 3669 (1970).
KROMPHARDT, H. , GROBECKER, H. , RING, K. , HEINZ, E. : Biochim. Biophys. Acta $\underline{74}$, 549 (1963).
POTASHNER, S. J. , JOHNSTONE, R. M. : Biochim. Biophys. Acta $\underline{233}$, 91 (1971).
REID, M. , EDDY, A.A.: Biochem. J. $\underline{124}$, 951 (1971).
RIGGS, T. R. , WALKER, L. M. , CHRISTENSEN, H. N. : J. biol. Chem. $\underline{233}$, 1479 (1958).
SCHAFER, J.A. , JACQUEZ, J.A. : Biochim. Biophys. Acta $\underline{135}$, 1081 (1967).
SCHULTZ, S. G. , CURRAN, P. F. : Physiol. Rev. $\underline{50}$, 637 (1970).
SIMONSEN, L.O. , NIELSEN, A. T. : Biochim. Biophys. Acta $\underline{241}$, 522 (1971).
VIDAVER, G.A.: Biochemistry $\underline{3}$, 803 (1964).
VIDAVER, G.A. , SHEPHERD, S. L. : J. Biol. Chem. $\underline{243}$, 6140 (1968).
WHEELER, K. P. , CHRISTENSEN, H. N. : J. biol. Chem. $\underline{242}$, 3782 (1967).
WILBRANDT, W. , ROSENBERG, T. : Pharmacol. Rev. $\underline{13}$, 109 (1961).

# Electrolyte Effects on the Transport of Cationic Amino Acids*

Halvor N. Christensen
Department of Biological Chemistry, The University of Michigan, Ann Arbor,
Michigan, USA

(Paper to occur in the introductory section of August 3)

I should like to begin by expressing my appreciation of the presentation just made
by Dr. Eddy. Through several years he has carefully focussed his attention on the
smaller contribution made by the $K^+$ gradient, along with the larger one made by
$Na^+$ gradient. His new results suggest that the $K^+$ gradient may exert a much larger
energizing effect when the ATP levels have not been depressed. As I heard the ion-
gradient hypothesis so beautifully and challengingly explained, I thought to myself,
this corresponds well to the possibility that Riggs and I saw in 1952 as a basis for
the $Na^+$ for $K^+$ exchange accompanying glycine uptake by the Ehrlich cell. We had
results even then that might have justified our limiting our consideration to the
sodium ion (see Fig. 1) as we developed the alkali-metal ion-gradient hypothesis,

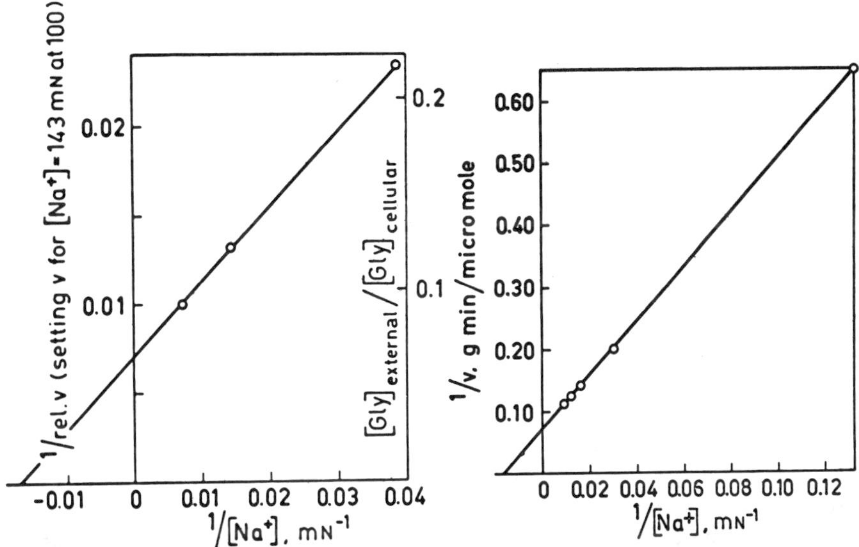

Fig. 1. Effect of external sodium ion concentration on the ability of the Ehrlich ascites tumor cell to accumulate glycine. Choline replaced $Na^+$ in the suspending medium. Left, uptake during 1 hr, observed in 1952 by Christensen and Riggs. Right,
uptake during 2 min observed in 1961 by Kromphardt et al. The half-saturating concentrations of $Na^+$ are very similar in the two cases. Reproduced with permission
from Membranes and Ion Transport, E. E. Bittar, ed., Wiley-Interscience, 1970,
v. 1, p. 365

* The experiments from our laboratory described here have been supported in
part by a grant (HD-01233) from the Institute for Child Health and Human Development, National Institutes of Health, U. S. P. H. S.

but there was accompanying evidence urging us to retain interest in the potassium ion. As a personal note and in a lighter vein I may mention a subjective reason: My distinguished mentor, Baird Hastings, has long had firm convictions about potassium, and he would never have forgiven me if we had overlooked the role of that ion. I must admit that from 1961 to about 1965 I came to be slightly rueful that we had not got the alternative forms of the alkali-metal gradient hypothesis into the titles and abstracts of our papers, as well as into the text. The subsequent development of the subject has, however, I am sure been rewarding to everyone.

Because of the excellent summary by Dr. Eddy, I may turn the attention of my presentation away from direct tests of the alkali-metal ion-gradient hypothesis to findings that reveal further complexities in the linkage between the fluxes of these ions and those of organic metabolites. I have been asked first to consider electrolyte effects on the transport of cationic amino acids.

First we may consider the inherent restraints placed on the uptake of cationic amino acids by their state of charge. Their net uptake must presumably be accompanied either by the displacement of cellular cations or by the uptake of an anion, for example $Cl^-$. On an electrogenic basis we do well to look for movements of $K^+$ or $Na^+$ out of the cell when lysine moves in. Whether the opposed fluxes are also coupled is of course a separate question.

The studies of Gale of lysine uptake by S. fecalis in 1947 led him to conclude that the movement into the cell is passive and probably occurs by diffusion, since $Q_{10}$ was only 1.4, uptake at $2^O$ being about one-fifth as fast as $38^O$. Accumulation occurred to the extent of 10- to 20-fold at $38^O$, which calls for a potential difference of 50 to 75 millivolts across the membrane if the uptake is indeed passive. Lysine showed only equivocal indications of saturating its own uptake at levels up to 15 mM in the external solution. The pH effect on lysine entry had a form to suggest that it is favored by a deprotonation with $pK' = 6.6$. That is, the degree of concentration of lysine achieved showed an increase with pH centering around pH 6.6. Gale attributed this pH effect to the formation of the isionic species as the permeant (2). To accept that interpretion, one has to suppose that $pK'_2$ of lysine, ordinarily 9.0, is strongly shifted in the course of the transport event, which could hardly occur for a transport by simple diffusion. The question is an important one: We are presently observing anomalies of transport reactivity that indicate shifts of apparent $pK'$s of basic amino acid by one or two pH units, a shift large enough to cause lysine to be transported to a considerable extent as a neutral amino acid, i.e., by neutral amino acid system. If the sidechain enters a highly apolar microenvironment at the receptor site, an amino group on that sidechain might be stabilized in its uncharged ($-NH_2$) form and hence be as acceptable as, for example, a hydroxyl group. For practical purposes this effect means that lysine and its analogs are not trustworthy as model basic amino acids. Arginine and arginine analogs are better because the sidechains are unambiguously cationic, $pK'_3$ being about thirteen for these guanidino amino acids; the synthetic models we use are in fact arginine analogs.

In any event lysine uptake by both S. fecalis and Staph. Aureus, paradoxically, was accompanied by $K^+$ uptake (2). The $Na^+$ content of the organisms did not change enough to explain this anomaly. A test for $Cl^-$ uptake was apparently not made. In contrast, Gale and his associates observed the loss of both $Na^+$ and $K^+$ from cells of Saccharomyces fragilis during lysine uptake. The release of accumulated lysine from S. fecalis could be prevented by 50 mM phosphate in the medium, and was affected by the composition of the amino acid mixture from which accumulation had accurred (4). Gale's results leave many fascinating unanswered questions. I wish microbiologists gave consideration more frequently to the effect of small concentrations of inorganic ions in their media. Eddy and Nowacki include lysine among the amino acids that occasion a linked shift of internal $K^+$ for external $H^+$ (5). Dr. Eddy may care to comment on the question of further ion shifs that might occur to produce electroneutrality.

<table>
<tr><td><u>$\gamma$-Zwitterion</u></td><td><u>Cation</u></td><td><u>$\alpha$-Zwitterion</u></td></tr>
</table>

$\gamma$-Zwitterion — $CH_2CH_2CH\text{-}COO^-$ with $NH_3^+$ and $NH_2$

Cation — $CH_2\text{-}CH_2\text{-}CH\text{-}COO^-$ with $NH_3^+$ and $NH_3^+$

$\alpha$-Zwitterion — $CH_2\text{-}CH_2\text{-}CH\text{-}COO^-$ with $NH_2$ and $NH_3^+$

|  |  |  |
|---|---|---|
| β system? | L$\overset{+}{y}$ system | <u>ASC</u> and <u>A</u> systems |

Found:

| partial inhibition by β-alanine | partial inhibition by lysine | All criteria for powerful uptake by System <u>A</u>; Negligible uptake by System <u>ASC</u> |
|---|---|---|

Conclusion: This amino acid reacts like an $\alpha$-Zwitterion with System <u>A</u>, but with System <u>ASC</u> it does not!

Fig. 2. Predicted routes of transport of $\alpha$, $\gamma$ -diaminobutyric acid for each of the ionic species likely to be present in neutral solution. See text for discussion

Almost two decades ago my associates and I were studying the uptake of a basic amino acid that is really ambiguous in its behavior, $\alpha$, $\gamma$ -diaminobutyric acid (Fig. 2). To system A of the Ehrlich cell it looks like a neutral amino acid substrate (6). To another $Na^+$-dependent system ASC it ought also to be a substrate to at least as great an extent; but it is not. Accordingly it is clear that one of these transport systems stabilizes the amino acid in the form of the $\alpha$ -zwitterion, but the other does not (3). In any event that portion of its uptake that occurs as a neutral amino acid is $Na^+$-dependent; the part that occurs as a basic amino acid is not. I wonder if a parallel phenomenon may not explain the partial $Na^+$ dependence of lysine transport by the isolated small intestine.

The uptake of diaminobutyric acid by cells is accompanied by wholesale displacement of $K^+$ and $Na^+$ from the cell, and by uptake of $Cl^-$ and hence of water to cause tremendous swelling. A feature that we did not at first understand appears in the 1958 result illustrated Fig. 3: When $Na^+$ displaced $K^+$ from the Ehrlich cell its capability for taking up glycine decreased. When the amino acid was the displacing cation, however, the capability for glycine uptake increased (7). Dr. Eddy saw the answer to this paradox, I think, in the circumstance that $Na^+$ as well as $K^+$ is displaced from the cell by this amino acid. The favorable effect of decreased cellular $Na^+$ was apparently more influential than the unfavorable effect of decreased $K^+$ in affecting net transport (Table 1).

An interesting, related phenomenon is seen in potassium deficiency in the rat. In this deficiency the cellular contents of arginine and lysine are substantially increased (8, 9). Perhaps this effect arises from the restraints that fall on the fluxes of these amino acids when the level of cations exchangeable with them is low. The re-

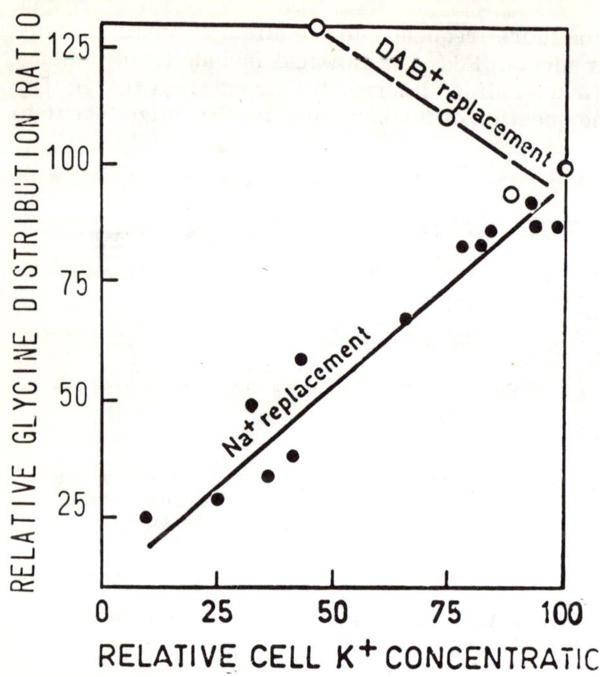



Fig. 3. Effect of displacement of cellular potassium on the rate of glycine uptake by the Ehrlich cell. Smaller circles, K$^+$ displaced by Na$^+$ through treatment of the cells with 2, 4-dinitrophenol. Larger circles, cellular K$^+$ and Na$^+$ extensively displaced by the organic cation, $\alpha$, $\gamma$-diaminobutyric acid, by incubation in the presence of this amino acid. See t ext for discussion. Reproduced with permission from Membranes and Ion Transport, E. E. Bittar, ed., Wiley-Interscience, 1970, v. 1, p. 365

sult recalls the situation in two invertebrate phyla, where it is not K$^+$ and Mg$^{++}$, but the amino acids, including anionic and cationic ones, that provide most of the osmotic pressure of tissues (10).

Fig. 4 shows two aspects of the action of Na$^+$ in the transport of basic amino acids by the Ehrlich cell. The exodus of previously accumulated homoarginine is faster into Na$^+$-KRB than into choline KRB (points along ordinate). External Na$^+$ is accordingly a stimulant to homoarginine exodus. But as we add homoarginine to the suspending medium, it proves much more effective than Na$^+$ in exchanging with internal homoarginine. Therefore external Na$^+$ changes from a weak stimulant to homoarginine exodus to a competitive inhibitor with external homoarginine for the exchange process.

Some of you know the somewhat drawn-out story of how we came to understand a rather more important role of the alkali metal ions in the uptake of basic amino acids. For five years we really didn't understand why some neutral amino acids could be such effective inhibitors of basic amino acid migration. We provisionally proposed, as others have, transport systems that really don't seem to care whether the substrate amino acid has a charged sidechain or not. Such systems seem inherently implausible to me, and they failed to explain some of our observations. Finally it came to us that an alkali-metal cation was needed to act as cosubstrate with the neutral amino acid, either to inhibit transport or stimulate countertransport of the basic amino acid. Furthermore, we came to see that the structure of the neutral amino acid determined how well it could cooperate with Na$^+$ to interact with the basic amino acid transport system. The structural requirements for inhibition of basic amino acid transport are quite different than those for transport by the neutral amino acid systems. Fig. 5 shows one aspect of the interaction, the cooperation of Na$^+$ and homoserine to inhibit the uptake of homoarginine by the Ehrlich cell. Fig. 6 shows another aspect: Na$^+$ becomes a much stronger stimulant of

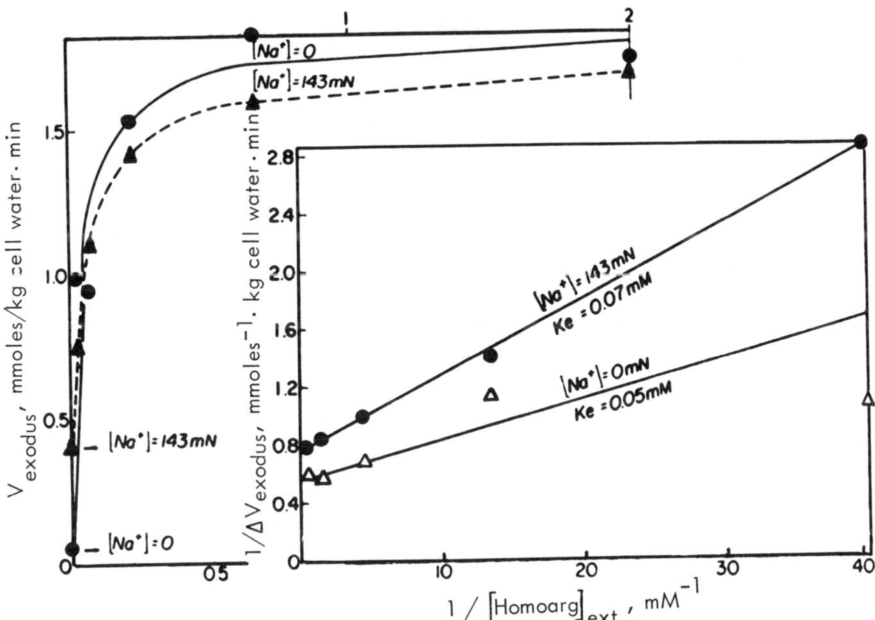



Fig. 4. Diminished stimulation of exodus of $^{14}$C-homoarginine by external homoarginine in the presence of Na⁺. The rates of exodus of previously accumulated
homoarginine actually observed are plotted at the left in the absence of Na⁺, at
various external homoarginine concentrations. Note that in absence of external
homoarginine, Na⁺ increases the rate of exodus five-fold. At the right, the augmentation of exodus due to external homoarginine is plotted according to Lineweaver
and Burk. For further details see reference 13. Reproduced with permission from
Christensen, Handlogten and Thomas, Proc. Nat. Acad. Sci. U. S. <u>63</u>, 948 (1969)

homoarginine exodus when homoserine or some other suitable neutral amino acid
is present. Although we have not been able to measure the stoichiometry precisely,
the ratio of exchange appears to be one Na⁺ plus one neutral amino acid molecule,
exchanging for one homoarginine molecule.

Fig. 7 shows our provisional, two-dimensional picture of the transport receptor site
in question. We were led to suppose that Na⁺ occupies the position otherwise taken
by the distal cationic structure on the basic amino acid. We do not yet know for sure
the three-dimensional geometry of this site.

This picture was reached from our ability to introduce reporter structures into the
amino acid sidechain that would sense the location of the sodium ion, as illustrated
in Fig. 8. Here the straight-chained amino acids of from 3 to 6 carbon atoms were
used. These are expecially  reactive with the basic amino acid system, a hydrocarbon
branch on the $\alpha$ or $\beta$ carbon atom being unfavorable. Similar selectivity may, in-
cidentally be noted in the interactions of neutral amino acids with lysine transport
by intestinal epithelial cells (11) and by pulmonary macrophages (12). When a hydroxyl group was introduced on carbon 5, a maximal enhancement of the reactivity of
the neutral amino acid resulted, as may be noted in the figure (13). Sulfhydryl groups
served also.

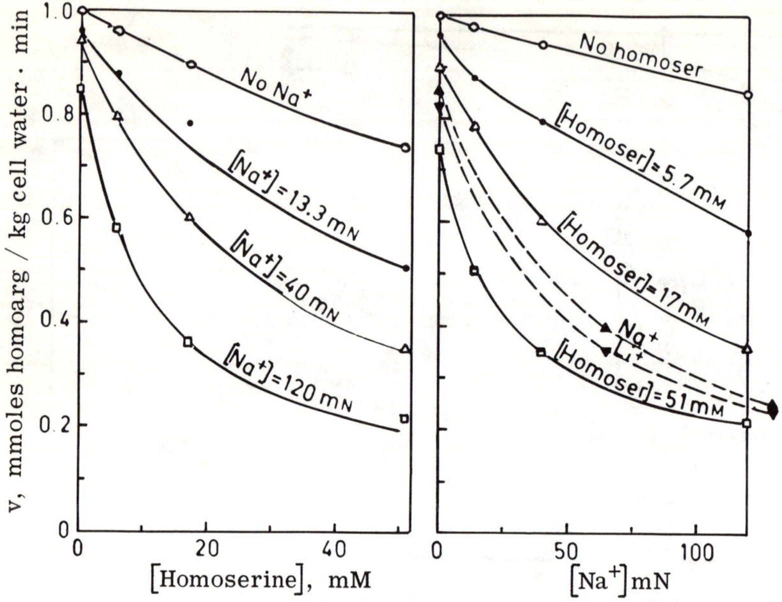



Fig. 5. Inhibitory effect of increasing concentrations of homoserine and Na⁺ on homoarginine uptake by the Ehrlich ascites tumor cell. Uptake measured during 1 min. See reference 13 for further details. Reproduced with permission from Christensen, Handlogten and Thomas, Proc. Nat. Acad. Sci. U. S. $\underline{63}$, 948 (1969)

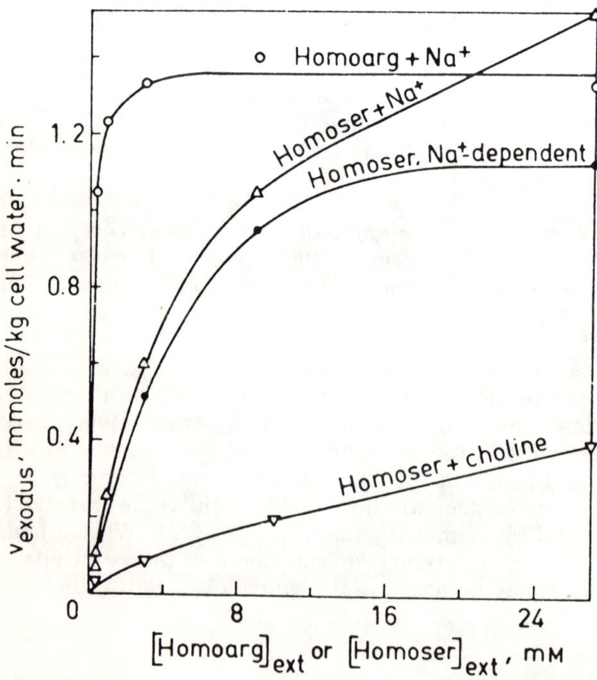



Fig. 6. Concentration dependence of the stimulation of $^{14}$C-homoarginine exodus from the Ehrlich cell by either external homoarginine or by external homoserine, at $[\text{Na}^+] = 0$ and 116 mN. The next to the lowest curve shows by difference the Na⁺-dependent stimulation by homoserine, which is half-maximal at 5 mM. Homoserine, in contrast, saturates its own uptake as a neutral amino acid at 0.2 mM also at a 0.12 M Na⁺ concentration. See reference 13 for further details. Reproduced with permission from Christensen, Handlogten and Thomas, Proc. Nat. Acad. Sci. U. S. $\underline{63}$, 948 (1969)

Fig. 7. Two-dimensional visualization proposed for the receptor site of transport system Ly⁺ occupied by lysine (above), and occupied by homoserine and Na⁺ (below). Reproduced with permission from Federation of European Biochemical Society Symposia $\underline{20}$, 81 (1970)

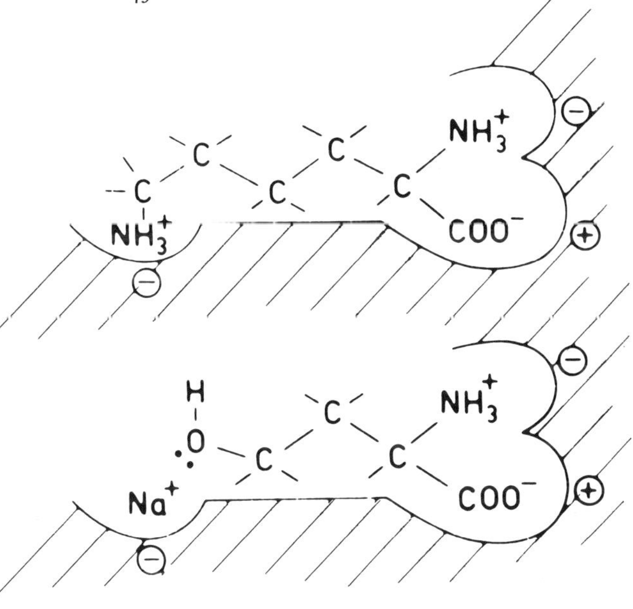


Another effect of getting the hydroxyl group into the optimal position may be illustrated. The selectivity among the alkali metal ions was ordinarily weak, falling in the expected series Li>Na>K>Rb>Cs. But when the hydroxyl group was placed on the appropriate carbon atom, a sharp upset occurred in this sequence (Fig. 9). The results shown here are for the same transport system for basic amino acids in the rabbit reticulocyte. In this cell the optimal position is at carbon 4. Note that Na⁺

Fig. 8. The effect of chain length and position of a hydroxyl group on the reactivity of neutral amino acids in inhibiting the uptake of $^{14}$C-homo-argenine by the Ehrlich cell (scale at right) or in stimulating its exodus (scale at left). The lower curve shows the results for the amino acids with apolar sidechains, $\alpha$-amino-n-butyric acid, norvaline and norleucine. Note that threonine is no more effective than the first of these and that methionine is no more effective than norleucine. The upper two curves are for the corresponding omega-hydroxy amino

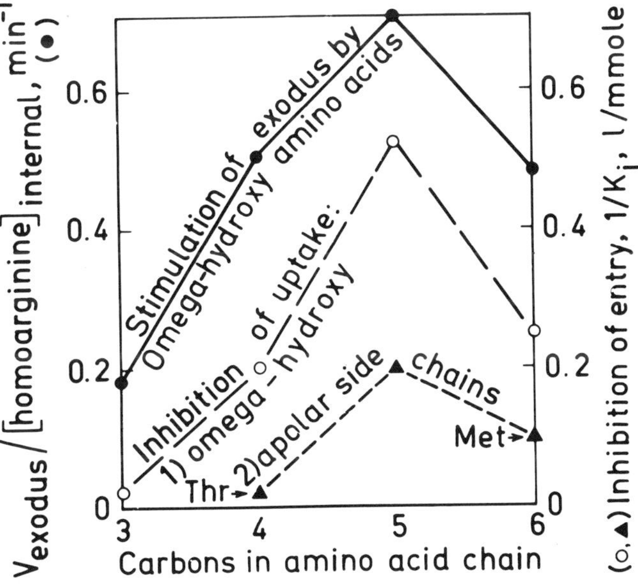


acids. Note how much more effective homoserine is than threonine. Results from the experiments of reference (13)

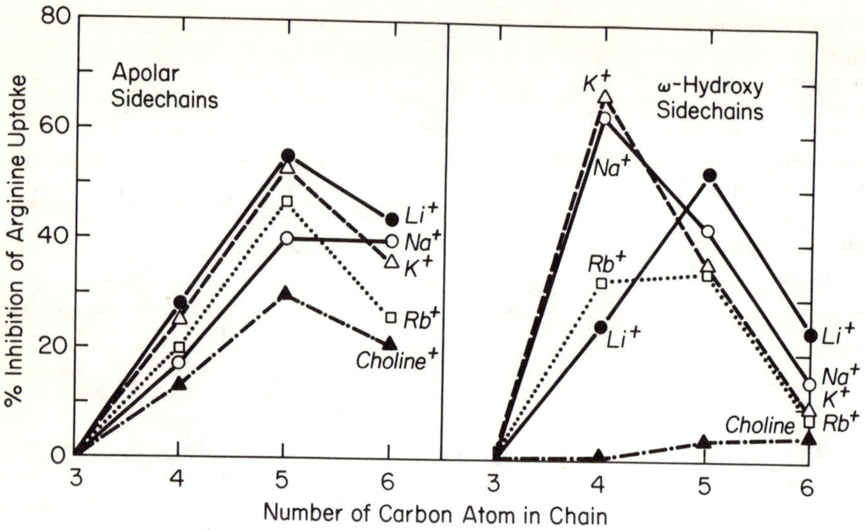



Fig. 9. Selectivity among alkali-metal ions in their cooperation with neutral amino acids to inhibit arginine uptake by the rabbit reticulocyte. Effect of linear chain length and of the presence of an omega-hydroxyl group on the neutral amino acid sidechain. The indicated alkali-metal ions were at 72 mN, replacing the corresponding amount of choline. The inhibitory amino acid was at 50 mM, also replacing choline. Labeled arginine was at 0.05 mM. See reference (14) for details. Reproduced with permission from Thomas, Shao and Christensen, J. Biol. Chem. 246, 1677 (1971)

and K+ emerge as far and away the most effective when the hydroxyl group is on carbon 4 (14). It is not strange that the uptake of arginine by this cell is associated with the exodus of neutral amino acids and K+.

Somewhat similar results were obtained for the Ehrlich cell, except in that case the optimal position of the hydroxyl group is on carbon 5, rather than on carbon 4 as in red blood cells. When the hydroxyl group was on carbon 5, the relative effectiveness of Na+ and Li+ was abruptly inverted. In this case, however, K+ did not join Na+ in this enhancement of reactivity. Rb+ and Cs+ had successively smaller effects (14).

Some of you will be aware that we have found a similar interaction in a neutral amino acid transport system, except in that case by far the preferred substrate is the neutral amino acid plus Na+, and the inferior reactant the cationic amino acid (15). Indeed the cationic amino acid will not migrate or exchange in that case; it merely acts as a competitive inhibitor. Let me emphasize that in other respects also System Ly+ and ASC prove themselves two different transport systems: the $K_i$ and $K_m$ values are totally wrong for any other view (13, 16). Furthermore, the cationic system survives the maturation of the rabbit reticulocyte, while the neutral system does not. In addition, the specificity of the neutral system to Na+ permits us to study the operation of the cationic system alone in the presence of Li+, measuring Li+ fluxes by atomic absorption spectrometry.

We have naturally wondered whether univalent cations might participate in the action of neutral amino acids to inhibit uptake of cationic amino acids by yeast and

Neurospora. We have obtained evidence that the alkali-metal ions do indeed enhance the inhibitory action of the neutral amino acids on arginine uptake by a Saccharomyces cerevisiae strain. We have not yet, however, discovered any structural feature of the amino acid that leads to any sharp selectivity among the alkali-metal ions (14). Hence we have no evidence in these microorganisms for what we believe is a most interesting conclusion to be drawn from the results with the animal cells, namely that $Na^+$ not only takes its place in juxtaposition to the neutral amino acid substrate at the recognition site, but also appears to form a bridge to it when the amino acid structure is favorable. That bridging accounts, we believe, for the sharp change in selectivity when the hydroxyl group on the amino acid side-chain is in the optimal position. From chelation chemistry we know that a most sensitive  determinant of selective coordination of a metal is the distance to be bridged. Perhaps the most telling evidence for such a bridging action comes from the interaction between $Na^+$ and the neutral amino acid in the system for neutral amino acids that we call ASC (for alanine, serine and cysteine). The introduction of a hydroxyl group into proline on carbons 3 or 4 causes a strong enhancement

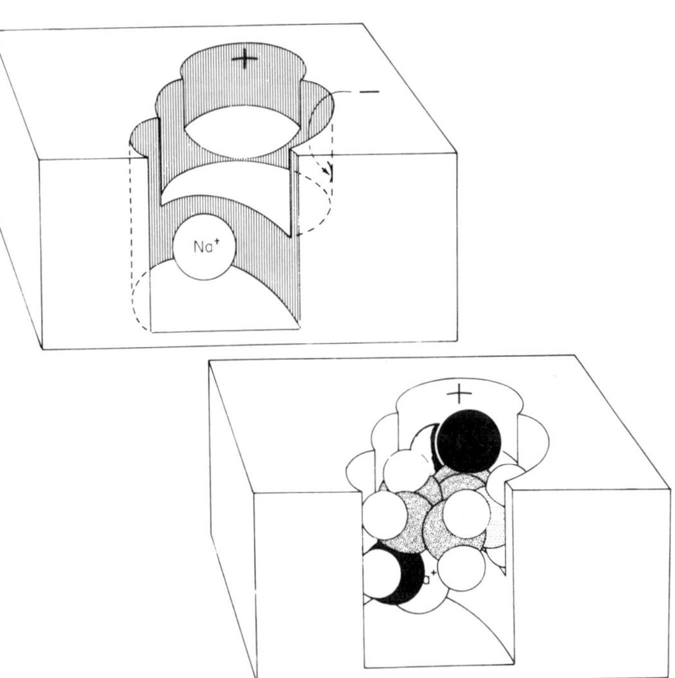


Fig. 10. Diagrammic representation of transport site ASC, showing $Na^+$ only (above) and both $Na^+$ and hydroxyproline (bottom) in suggested positions. No implication is intended that $Na^+$ must enter the site first. See text for discussion. Reproduced with permission from Thomas and Christensen, Biochem. Biophys. Res. Comm. <u>40</u>, 277 (1970)

in uptake, and in the interaction with $Na^+$ for uptake. But this effect occurs only if the hydroxyl group is trans to the carboxyl group. In the cis orientation, the hydroxyl group has only an unfavorable effect on the interaction. We were led therefore to localize the position taken by $Na^+$ somewhat as pictured diagrammatically in Fig. 10 (15). Not only the position, but the orientation of the hydroxyl group, as shown with the hydroxyprolines, is decisive to the interaction between the two cosubstrates.

Corresponding results are now being obtained with the principal $Na^+$-dependent (A) system of the Ehrlich cell, that is, with the transport system corresponding to the one discussed by Dr. Eddy, also to be discussed by Doctors Jacquez and Schafer, and by Doctors Schafer and Heinz. The signal provided by certain structures when they are close to  the position taken by $Na^+$ in that case is a changed selectivity between $Na^+$ and $Li^+$ (17). Table II above summarizes the evidence obtained for the position taken by $Na^+$ in these two transport systems.

<u>References</u>


1. CHRISTENSEN, H. N. , RIGGS, T. R. : J. Biol. Chem. <u>194</u>, 57 (1952).
2. GALE, E. F. : J. Gen. Microbiol. <u>1</u>, 53 (1947).
3. CHRISTENSEN, H. N. , HANDLOGTEN, M. E. , CULLEN, A. M. : Fed. Proc. <u>30</u>, 1116 (1971).
4. DAVIES, R. , FOULKES, J. P. , GALE, E. F. , BIGGER, L. C. : Biochem. J. <u>54</u>, 430 (1953).
5. EDDY, A. A. , NOWACKI, J. A. : Biochem. J. <u>122</u>, 701 (1971).
6. CHRISTENSEN, H. N. , LIANG, M. : J. Biol. Chem. <u>241</u>, 5542 (1966).
7. RIGGS, T. R. , WALKER, L. M. , CHRISTENSEN, H. N. : J. Biol. Chem. <u>233</u>, 1479 (1958).
8. ECKEL, R. E. , POPE, C. E. , II, NORRIS, J. E. C. : Arch. Biochem. Biophys. <u>52</u>, 293 (1954).
9. IACOBELLES, M. E. , MANTWYLER, E. , DODGEN, C. L. : Am. J. Physiol. <u>185</u>, 275 (1956).
10. CAMIEN, M. N. , SARLET, H. , DUCHATEAN, G. , FLORKIN, M. : J. Biol. Chem. 193, 881 (1951).
11. REISER, S. , CHRISTIANSEN, P. A. : Biochim. Biophys. Acta 241, 102 (1971).
12. TSAN, M. -F. , BERLIN, R. D. : Biochim. Biophys. Acta <u>241</u>, 155 (1971).
13. CHRISTENSEN, H. N. , HANDLOGTEN, M. E. , THOMAS, E. L. : Proc. Nat. Acad. Sci. U. S. 63, 948 (1969).
14. THOMAS, E. L. , SHAO, T. -C. , CHRISTENSEN, H. N. : J. Biol. Chem. <u>246</u>, 1677 (1971).
15. THOMAS, E. L. , CHRISTENSEN, H. N. : J. Biol. Chem. 246, 1682 (1971).
16. ANTONIOLI, J. A. , CHRISTENSEN, H. N. : J. Biol. Chem. <u>244</u>, 1505 (1969).
17. Unpublished results, CHRISTENSEN, H. N. , HANDLOGTEN, M. E.
18. KOSER, B. H. , CHRISTENSEN, H. N. : Biochim. Biophys. Acta <u>241</u>, 9 (1971).

TABLE I
The several transport systems closely relevant to this paper are described here:

| Designation | Formally defined for | Characterizing substrate | Some preferred natural substrates | Largely excluded amino acids, besides dicarboxylic | Na$^+$ dependence | Exchanging properties | Other occurrences |
|---|---|---|---|---|---|---|---|
| L | Ehrlich cell | 2-aminonorbornane-2-carboxylic acid (BCH) | leu, phe, met, isoleu, norleu | N-methyl & cationic amino acids; probably AIB[a] | none detected | strong | ubiquitous; mature mammalian erythrocyte (compare System LIV of E. coli) |
| A | Ehrlich cell | MeAIB N-methyl-ala | ser, pro, gly, met, sarcosine | BCH, cationic amino acids[a] | first-order[b] (Li$^+$ serves) | weak | ubiquitous, but conspicuously absent in erythrocytes[c] |
| ASC | Ehrlich cell | A substrate restricted to this system is unknown | 3- to 5-carbon aliphatic, hydroxy-aliphatic amino acids; cysteine; proline | N-methyl & cationic amino acids[d] | first-order[e]. No other alkali.-metal serves; Li$^+$ inhibits | variable; low in Ehrlich cell | ubiquitous |
| L$_y^+$ | Ehrlich cell; red blood cells | homoarginine; 4-amino-1-guanyl-piperidine-4-carboxylic acid | arginine lysine etc. (ordinarily not cysteine) | neutral amino acids, in absence of alkali-metal ions | (asymmetric; see text) (Li$^+$ serves) | strong | ubiquitous; similar system in many microorganisms |

[a] Most natural, neutral amino acids are to some degree reactive.

[b] pH sensitivity is high.

[c] System Gly of kidney, intestine and red blood cells may well be a more restrictive, differentiative variant of ASC.

[d] Methionine and norleucine are strong inhibitors but show no transport. System is unusually stereospecific.

[e] Flux stoichiometry of cotransport variable.

## TABLE II

Evidence of the nature of the interaction between $Na^+$ and neutral amino acids in transport systems ASC of the rabbit reticulocyte and the pigeon red cell, and in System $Ly^+$ of these cells and the Ehrlich cell.

1. Competition between (neutral amino acid + $Na^+$) and (cationic amino acid without $Na^+$) seen in both systems.

2. The guanidine and methylguanidine cations are more reactive than $NH_4^+$, in association with neutral amino acid, corresponding to the greater reactivity of arginine than of lysine.

3. A hydroxyl group on a specific carbon atom causes mutual enhancement of the transport reactivity of the neutral amino acid and of the $Na^+$.

4. When the hydroxyl group is on the right carbon atom (carbon five of the Ehrlich cell; carbon four of the red blood cells) selectivity among alkali-metal ions is sharply modified.

5. Flux augmentation ratio, $Na^+$ to neutral amino acid, is sharply increased in System ASC of the pigeon red blood cell when the hydroxyl group is on carbon four (17).

6. The hydroxyl group of hydroxyproline must be trans to the carboxyl group for effects three to five.

# Transport of Amino Acids in Ehrlich Ascites Cells and Mouse Pancreas

Evidence against the $Na^+$ or alkali metal gradient hypothesis

R. M. Johnstone, Ph. D.
Department of Biochemistry, McGill University, Montreal, Quebec, Canada

The proposal that the $Na^+$ (or alkali metal cation) gradient(s) provide all the necessary motive force for accumulation of organic compounds against their respective concentration gradients has received wide support in several experimental systems[1-11].

If one considers the question of the theoretical availability of energy from the ion gradients (and it is assumed that the intracellular ionic activities are nearly equal to the measured concentrations), it is clear that under the normal circumstances found in mammalian cells, the theoretical energy available from the $Na^+$ gradient alone is frequently more than enough to provide adequate energy for accumulation of electrically neutral molecules if a 1:1 relationship prevails between ion and organic solute transport. If the sum of the energies from the $Na^+$ and $K^+$ gradient is considered, then the excess of potential energy available is several times greater than that required for accumulation of organic solutes in most mammalian systems. Thus a relationship between the energy available from cation gradients and accumulation of organic solutes will be seen experimentally only when the cation gradients are considerably reduced.

Many of the studies on the relationship between ion gradients and organic solute accumulation have been carried out under conditions of lowered extracellular $Na^+$ where the system is operating at a fraction of its optimal activity. Our approach to the question of the energy source for accumulation of amino acids in Ehrlich ascites cells and in mouse pancreas has been two fold (12, 13, 14). Firstly, we have asked whether under conditions of low cellular ATP (less than O. 1 mM) and near normal cation distributions, (normal meaning that the concentrations of cations found intra and extracellularly are those found in respiring cells incubated in 145 mM $Na^+$, 5. 7 mM $K^+$ and 154 mM $Cl^-$, and some suitable buffer), accumulation of amino acids is comparable to that abtained in cells containing high ATP levels (2-4 mM). For example, cells, high and low in ATP, are compared at equal or near equal cation distributions. Unfortunately, with the Ehrlich ascites cells (as with most cells) in absence of cellular ATP, the cellular cation distributions tend to approach those of the medium fairly rapidly, especially with respect to sodium.

However, if experimental periods of 30 minutes are used, asymmetric distributions of the cations can be maintained and one can then compare amino acid accumulation in presence and absence of cellular ATP at nearly comparable cation distributions. The results of these experiments, shown in Table 1, indicate that cation gradients will support accumulation of amino acids against their respective concentration gradients. The direction of amino acid movement is determined by the direction of the $Na^+$ and $K^+$ gradients and occurs in the direction of flow of $Na^+$ and opposite to that of $K^+$. Glycine is transported to a greater extent than methionine which in turn is greater than leucine. The relative transport activities are consistent with the known behavior of these amino acids in ATP-containing Ehrlich ascites cells (15). Thus we conclude that the $Na^+$ and $K^+$ gradients can bring about some accumulation of amino acids against a concentration gradient. However, when the accumulation of amino acids in cells depleted of ATP is compared with that in cells containing ATP, there is a 10-fold difference in accumulation (Table 2) for cation gradients of similar magnitude. Hense, it does not appear from these data, that the cation gradients provide sufficient energy for normal rates of accumulation of amino acids[*].

The technical difficulties in maintaining cation gradients in cells depleted of ATP require that the conclusions drawn be substantiated using another experimental approach. Therefore, in our second approach, we manipulated the $Na^+$ (and $K^+$) gradients in cells containing ATP by elevating intracellular $Na^+$. If the $Na^+$ gradient hypothesis is correct, certain corollaries should obtain.

At fixed extracellular concentrations of $Na^+$ and $K^+$, accumulation of organic solutes at steady state should be related to the magnitude of the $Na^+$ (or $Na^+$ + $K^+$) gradient. Comparison of gradients of similar magnitude but at differing extracellular cation concentrations is avoided to prevent any confusion of an effect of extracellular cation concentration with that of the gradient.

The results in Tables 3 and 4 show that in Ehrlich cells a high intracellular $Na^+$ does not impair the accumulation of glycine or methionine at steady state. It should be noted in this context that (i) the $Na^+$ gradient is almost absent and (ii) the $K^+$ gradient is substantially reduced. It is known that the sum of $\left[Na^+\right]$ + $\left[K^+\right]$ is about 200 mM in these cells irrespective of the conditions of incubation. Elevation of cellular $Na^+$ is therefore accompanied by a decrease in cellular $K^+$. The data in Table 3, (comparing lines 1 and 3), show that reduction of the $Na^+$ gradient by nearly 100% and the $K^+$ gradient by 30% has little effect on the accumulation of glycine provided that cellular ATP is maintained. However, if cellular ATP is decreased, uptake of both glycine and methionine is reduced despite the fact that the ion gradients are not appreciably altered (compare lines 3 and 5), Table 3). These data are therefore inconsistent with the gradient hypothesis whether one considers the $Na^+$ gradient alone or the $Na^+$ + $K^+$ gradients.

A second prediction of the $Na^+$ gradient hypothesis is that the movement of the organic solute must occur along the $Na^+$ gradient. The data in Fig. 1 however, show that accumulation of amino acids will continue despite a reversed $Na^+$ gradient. Only when the extracellular $Na^+$ is very low (20 mM or less) and internal $Na^+$ high, is there a tendency for the organic solute to move out of the cell against a concentration gradient. Thus to reverse the direction of movement of organic solute, extracellular $Na^+$ must be low in relation to intracellular $Na^+$. If the extracellular $Na^+$ is 80 mM and about 50% lower than intracellular $Na^+$, movement of organic solute into the cell against a concentration gradient will continue to take place in the Ehrlich cells containing ATP. Hence two conditions are necessary to reverse the direction of flow: (a) low extracellular concentration of sodium to reduce the normal inward activity and (b) high intracellular $Na^+$ to produce a reversal in the direction of sodium movement.

---

[*] In a recent communication by J. A. Schafer and E. Heinz, Biochim. Biophys. Acta <u>249</u>, 15 (1971), it was also concluded that an energy source other than the cation gradients provided some of the energy for amino acid uptake.

Fig. 1. The cells were pre-incubated for 1 h at $10^O$ in normal Krebs-Ringer or at $37^O$ in Krebs-Ringer containing $10^{-4}$M 2, 4-dinitrophenol. The medium contained 3 mM L-Me-$^{14}$C-methionine, specific activity 120 counts/min per nmole. This gave rise to cells which contained nearly equal $Na^+$ (ca. 150 mM) and methionine levels but greatly different ATP levels. One lot of cells from each group was transferred to media containing 20 mM $Na^+$ or 80 mM $Na^+$ and the same concentration and specific activity of methionine as that found in the cells after preincubation. $Na^+$ was replaced by choline. Changes in$^{14}$C methionine concentration were measured over a period of 30 min at $25^O$. ●-●, ATP and 80 mM $Na^+$; ▲-▲ , 2, 4-dinitrophenol and 80 mM $Na^+$; o·····o, ATP and 20 mM $Na^+$; Δ····Δ , 2, 4-dinitrophenol and 20 mM $Na^+$

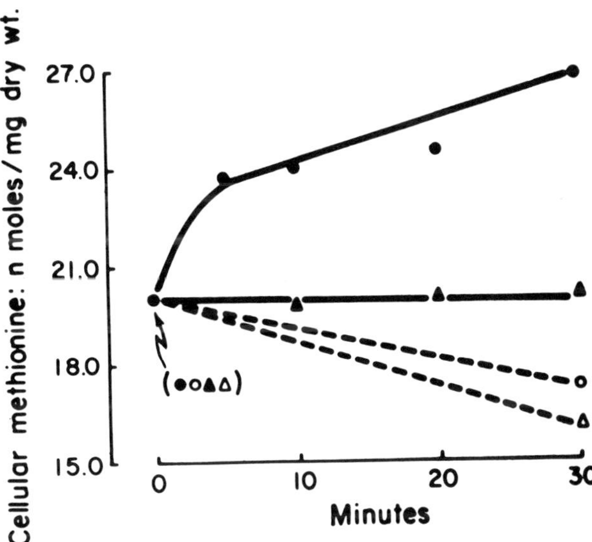


The third inconsistency with the $Na^+$ gradient hypothesis is shown by experiments on the initial rate of amino acid uptake. The hypothesis assumes a reversible carrier system where the kinetic constants are modified by the cations found on each side of the membrane. The asymmetric distribution of the cations ensures that the kinetic constants in the two directions are different so that influx exceeds efflux over a considerable range of solute concentrations, thus permitting accumulation. However, no role is assigned to the $Na^+$ or $K^+$ gradients (or ATP to maintain the cation gradients) until the intracellular concentration of organic solute approaches that of the medium. The role of the gradient in accumulation, according to the hypothesis, is to lower the efflux for a given concentration of organic solute. So long as the intracellular concentration of organic solute is less than that in the extracellular medium, there is no need for the gradient or the energy from the gradient, the reaction being downhill. In our experimental system, we have been able to demonstrate that the initial rate of amino acid uptake before the intracellular amino acid concentration has reached that in the medium, is a process stimulated by the availability of metabolic energy. The data suggest that even the rate of downhill movement of amino acids is dependent on cellular ATP (Fig. 2). Thus overall, these data drawn from the Ehrlich cell system are inconsistent with the predictions of the $Na^+$ gradient hypothesis.

Turning to our experimental system, the mouse pancreas, we find again that much of our data are contradictory to the predictions of $Na^+$ and $K^+$ gradient hypothsis. The mouse pancreas is capable of very extensive accumulation of amino acids, of which glycine is a good example (16, 17). Under in vitro conditions of incubation, however, the pancreas does not maintain a $Na^+$ gradient

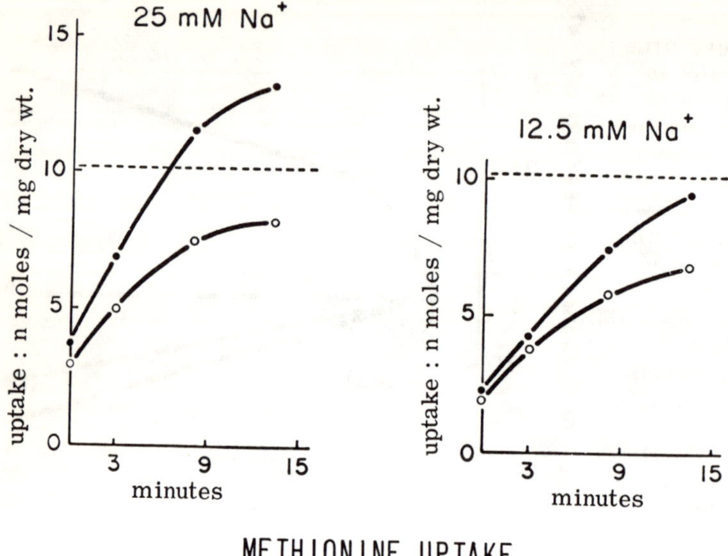



## METHIONINE UPTAKE

Fig. 2. Cells were preincubated at 37° for 1 h in modified Ringer solutions containing $10^{-4}$ M 2, 4-dinitrophenol in which $K^+$ replaced $Na^+$. This treatment gave rise to cells with near normal $K^+$ levels and little ATP. After preincubation, the cells were divided into two lots. To one, were added $10^{-4}$ M 2, 4-dinitrophenol and 10 mM pyruvate. To the other were added 1 mM glucose and 10 mM pyruvate but no 2, 4-dinitrophenol. The media contained the $Na^+$ concentrations shown at the top of the figure and isotonicity was maintained with choline chloride; $K^+$ was kept at 8 mM. The cells were incubated for 2 min at 25° and then 2. 5 mM L-Me-$^{14}$C methionine, specific activity 180 counts/min per nmole, was added and uptake of radioactivity measured. ● - ●, glucose, no 2, 4-dinitrophenol; o-o-o, 2, 4-dinitrophenol. The dashed lines indicate the position where the intracellular and extracellular concentrations become equal

(Table 5). Although anaerobiosis and DNP cause similar changes in cation distribution in this tissue, glycine accumulation is twice as great in DNP as under $N_2$. Removal of extracellular $K^+$ also reduces accumulation and leads to a further small increase in cell $Na^+$.

Further evidence that the alkali metal gradients do not provide energy comes from experiments with elevated external $K^+$. The data show that in pancreas, increasing extracellular $K^+$ does not reduce amino acid accumulation (Table 6). For example, with extracellular $Na^+$ at 48 mM and $K^+$ at 7. 8 or 105 mM, similar accumulation is obtained. Since this tissue does not maintain a $Na^+$ gradient, the theoretical gradient energy at elevated extracellular $K^+$ is insufficient to account for glycine accumulation greater than 1. 7 times the medium concentration of glycine if a 1:1 relationship prevails between glycine and ion movements. With elevated $K^+$, glycine accumulation is more than 4 times that of the medium (Table 6, Exp. 1).

The action of ouabain on amino acid transport in the pancreas is also noteworthy in this regard. One of the frequent arguments used to substantive the view that movement of organic solutes is secondary to the pumping of $Na^+$ is the fact that there is a lag in the action of ouabain on organic solute transport, in contrast to its immediate effect on cations (4, 5). The interpretation of this finding has been that there is no direct effect of ouabain on organic solute transfer. Inhibition of organic solute transfer is presumed to be due to a dissipation of the ion gradients which then leads to an inhibition of organic solute transfer.

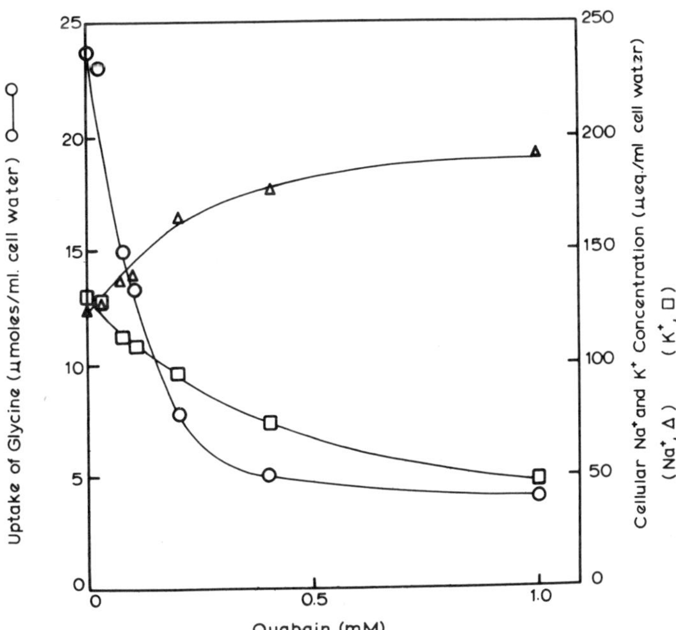



Fig. 3. Action of ouabain on glycine uptake and on $Na^+$ and $K^+$ distributions in mouse pancreas. The pancreas was incubated in standard medium containing 2 mM $^{14}C$-glycine with varying ouabain concentrations for 60 min. A 60 min incubation period was used, sothat steadystate conditions were examined. o-o, total glycine uptake; Δ-Δ, cell $Na^+$; □-□-, cell $K^+$. All values corrected for extracellular space and the concentration of solute therein

The data in Fig. 3 show, however, that in pancreas, ouabain has an effect on the accumulation of glycine at concentrations which have little measurable effect on their intracellular cation concentrations. In fact, it takes 10 times more ouabain to reduce cell $K^+$ by 50% than to reduce glycine uptake by 50%. It may be noted that at all ouabain concentrations tested there is some, albeit small, effect on the cation distributions and that the concentrations of ouabain used are those frequently used with rodent tissues to inhibit cation transport. The fact that amino acid accumulation at steady state is inhibited without a corresponding degree of modification of the ion gradients suggests that the action of ouabain on amino acid uptake is unlikely to occur as a result of altering ion gradients. The relative insensitivity of the cation distribution to ouabain may be partly due to the fact that in vitro handling has increased "leakiness" to ions, particularly $Na^+$ so that the additional inhibitory effects of ouabain are masked.

We have also obtained some evidence in both Ehrlich cells and pancreas that accumulation of amino acids is a function of ATP concentration or some related component. In Tables 3 and 4 it may be seen that in the Ehrlich cells, for gradients of nearly equal magnitude and with extracellular cations at the same concentrations, uptake of methionine and glycine follows the cellular ATP level (compare lines 3, 5 and 6 in Table 3; lines 2, 3 and4 in Table 4).

If cells are exposed to dinitrophenol and then transferred to fresh medium, free from dinitrophenol, there is a gradual restoration of the ATP level. If the dinitrophenol level is maintained, no restoration of ATP is observed. With and without DNP, these cells do not re-establish their cation gradients substantially within 30 min. (lines 5 and 6 in Table 3). However, either with glycine or methionine there is a  measurable difference in amino acid uptake when dinitrophenol is removed (compare bottom 2 lines in Tables 3 and 4).

In pancreas, the cation concentrations in presence or absence of dinitrophenol have reached a steady state after 15-30 min. incubation and the values at 60 min. have been chosen for display in Table 5. If the uptake of glycine were dependent on the cation gradients and the action of DNP on glycine uptake were a consequence of its action to reduce the ion gradients by virtue of the decreased $Na^+$ pump activity, exposure to DNP for periods longer than 30 min. should not decrease the uptake of amino acids beyond the decrease seen at 30 min. The data in Fig. 4 show that increasing time of exposure to dinitrophenol for periods up to two hours results in a steadily decreasing uptake of glycine. (Presence or absence of ATP has no effect on the efflux of glycine, Fig. 5). Furthermore, under aerobic conditions, glutamine or glutamic acid protect against the reduction of transport caused by DNP pretreatment (Table 7). Glutamine does not protect against the loss of transport activity caused by anaeribiosis. Citrate, pyruvate and glucose do not protect against the action of dinitrophenol on amino acid uptake. (In this context, it should be noted that in pancreas which has not first been incubated with dinitrophenol, the addition of glutamine does not increase glycine uptake).

There does not appear to be significant exchange between glycine and glutamine since in tissue first incubated with $^{14}C$-glutamine and DNP followed by incubation with $^{12}C$-glycine, there is no associated increase in the loss of $^{14}C$-glutamine from the cells as expected for exchange. Moreover, the increased uptake of glycine is more than twice the cellular level of $^{14}C$-glutamine.

These data are consistent with the conclusion that substrate level phosphorylation obtained from aerobic oxidation of glutamine or glutamate, via $\alpha$-ketoglutarate and succinyl CoA provides a low level of ATP even in presence of DNP whereas neither pyruvate nor citrate is capable of so doing. The lack of effect of glucose in this regard is also interesting since we have not been able to detect any glycolysis in the pancreas. Furthermore, it is known that pancreas has high endogenous levels of glutamate[18] which would account for the required two hour exposure to dinitrophenol to reduce glycine transport to minimal levels.

Additional evidence in support of direct participation of ATP in organic solute transport comes from experiments designed to measure the kinetic constants. The data in Table 5 have shown that there are modest differences in cellular cation concentrations in presence and absence of dinitrophenol. Determination of the $K_m$ and $V_{max}$ for transport (Fig. 6) show that both the $K_m$ and $V_{max}$ for glycine transport are altered in pancreas depleted of ATP. The major effect of ATP lack is to reduce the $V_{max}$ (as is also the effect of reduced extracellular $Na^+$) from 6.4 to 0.96 $\mu$ moles per gm dry wt/min. The $K_m$ is decreased from 6.7 to 2.5 mM by ATP lack. In contrast, in Ehrlich cells both lack of $Na^+$ or ATP results in an increase in the apparent $K_m$ values for glycine and methionine (13).

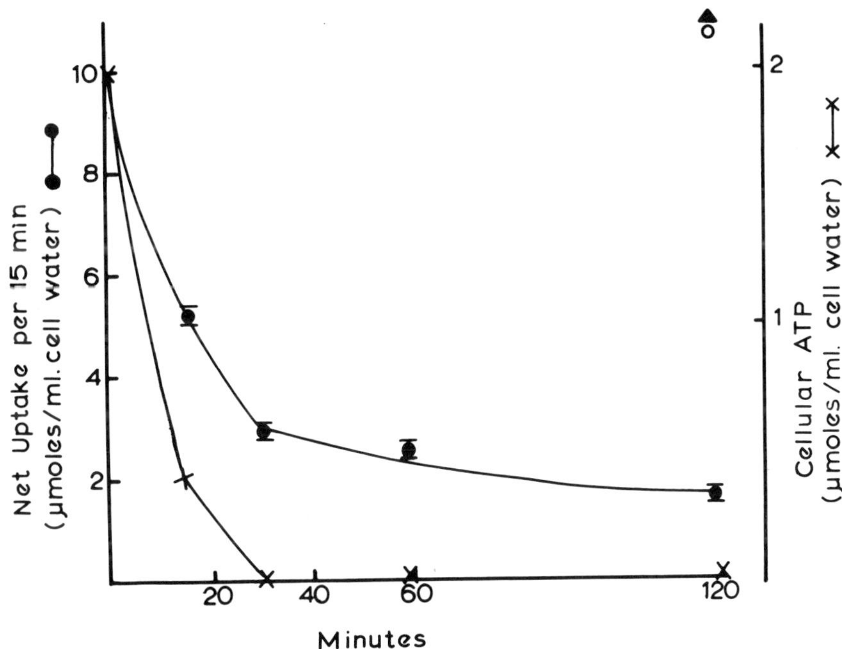

Fig. 4. Decrease in initial rate of glycine uptake with time of exposure to dinitrophenol in mouse pancreas. ATP concentration in a representative portion of freshly excited pancreas was estimated as well as in portions of the tissue that were exposed to dinitrophenol for 15, 30, 60 and 120 min, respectively. Portions of tissue were also incubated for 120 min without dinitrophenol to determine the maintenance of the cellular ATP level in the control situation (o). The uptake of glycine was measured for a period of 15 min. To measure the effect of pretreatment with dinitrophenol on the uptake of glycine, the following procedure was adopted: The tissue was preincubated with dinitrophenol for 0, 15, 30 and 120 min, after which 1-$^{14}$C glycine was added and uptake allowed to continue for a period of 15 min. A control preincubated without dinitrophenol for 120 min, was also included (▲). Thus, in this figure, glycine uptake after preincubation with dinitrophenol for 15 min actually represents tissue that was exposed to dinitrophenol and glycine simultaneously for 15 min whereas the ATP level is that found at the end of the 15 min period of exposure to dinitrophenol-● - ●, glycine uptake after preincubation with dinitrophenol; ▲ , glycine uptake after 120 min preincubation without dinitrophenol; x - x, ATP level during dinitrophenol treatment; O, ATP level after 120 min incubation with normal Krebs Ringer medium. All values are means of three or four closely agreeing determinations the spread of which are given for glycine. Values for glycine are net uptakes representing $A_c$ - $A_f$ (cellular concentration less medium concentration).

Thus, in addition to data that accumulation can occur (i) in absence of a Na$^+$ gradient or Na$^+$ + K$^+$ gradients (ii) in the direction opposite to the existing Na$^+$ gradient (iii) that accumulation can be obtained in excess of the theoretical gradient energy, we have also obtained some direct evidence that ATP is required for accumulation of amino acids. This evidence is: (1) For ion gradients of nearly equal magnitude and at equal extracellular cation concentrations, accumulation follows the cellular ATP level. (2) When cellular ATP is reduced, restoration of accumulation of amino acids follows restoration of the ATP level. (3) Experimentally,

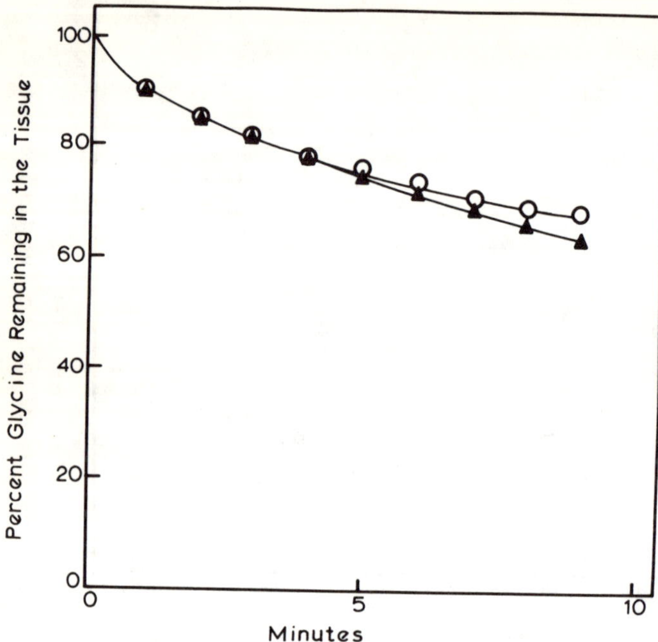

Fig. 5. Action of dinitrophenol on initial rate of glycine exodus in mouse pancreas. The tissue was preincubated with or without $10^{-4}$M dinitrophenol for 30 min and then $1\text{-}^{14}C$ glycine was added to give a final concentration of 2 mM (control) and of 8 mM (with dinitrophenol), respectively, in a final volume of 3 ml. The preincubation is essential since the cellular ATP level falls in the first 30 min of the dinitrophenol treatment (see Fig. 4). After the addition of glycine, the incubation was continued for another 60 min. At the end of this period, the cellular glycine content was nearly equal (20 mM) in both cases. To circumvent the problems of unequal extracellular glycine concentrations the following procedure was adopted. At the end of the incubation, the tissue was transferred to tared tubes, spun down, drained and weighed. Then 10 ml of normal Krebs Ringer buffer solution was added to each tube, stirred quickly for 5 sec, spun down again and drained. A portion of this washing medium was counted and it showed more than enough radioactivity to account for the extracellular space. Each tissue was transferred to 9 ml of fresh medium devoid of any amino acid but with and without dinitrophenol as in the preincubation. At various time intervals, 200 $\mu$l of the medium were removed and counted. After the final sampling, the radioactivity remaining in the tissue and medium were determined. The total radioactivity associated with the tissue at 0 time of incubation was calculated and used as the 100% control. o-o-o, control preincubation with 2 mM glycine; ▲-▲-▲-, preincubation with 8 mM glycine plus DNP

ion gradients alone, in ATP depleted cells, do not bring about adequate accumulation of amino acids to account for the accumulation observed under more physiological conditions. (4) The time required to abolish glycine accumulation with dinitrophenol in the pancreas is much greater than the time required to achieve a new steady state for the cellular cations and consistent with the conclusion that another energy source is maintaining glycine transport. (5) Glutamine and glutamate, not citrate or pyruvate, protect against reduction of transport of amino acids caused by dinitrophenol in the pancreas. (6) The kinetic constants for amino acid uptake in Ehrlich cells and in the pancreas are altered by cellular ATP above the alterations brought about by $Na^+$.

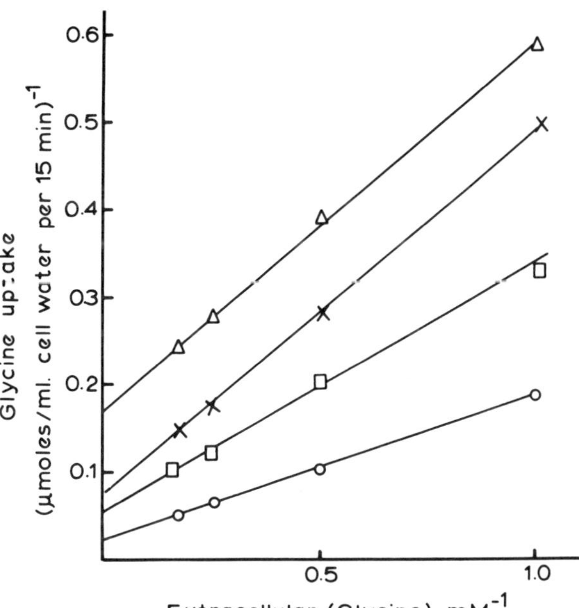



Fig. 6. Action of ATP and Na$^+$ on the kinetic constants for glycine transport. The pancreas was preincubated for 60 min in normal Krebs Ringer (o-o-o); 77 mM Na$^+$ ($\square$-$\square$-$\square$); 38 mM Na$^+$ (x-x-x); and normal Krebs Ringer containing $10^{-4}$M DNP ($\triangle$-$\triangle$-$\triangle$). NaCl was replaced by choline chloride where necessary. After preincubation, sufficient 1-$^{14}$C-glycine was added to each flask to attain the concentrations cited and incubation continued for 15 min. Each point is the average of 4-8 individual determinations in which the average value is less than $\pm$ 5% from the extreme value. The values for uptake represent net uptake, that is $A_c$ - $A_f$.

## References


1. CRANE, R. K.: Fed. Proc. 24, 1000 (1965).
2. CRANE, R. K., FORSTNER, G., WICHHOLZ, A.: Biochim. Biophys. Acta 109, 567 (1965).
3. SCHULTZ, S. G., in R. M. Dowben, Biological Membranes, Little, Brown and Co., Boston, Mass. 1969, p. 59.
4. CHEZ, R. A., PALMER, R. R., SCHULTZ, S. G., CURRAN, P. F.: J. Gen. Physiol. 50, 2357 (1967).
5. SCHULTZ, S. G., CURRAN, P. F.: Physiologist 12, 437 (1969).
6. VIDAVER, G. A.: Biochemistry 3, 803 (1964).
7. VIDAVER, G. A., SHEPHERD, S. L.: J. Biol. Chem. 243, 6140 (1968).
8. EDDY, A. A.: Biochem. J. 108, 489 (1968).
9. EDDY, A. A., MULCAHY, M. F., THOMSON, P. J.: Biochem. J. 103, 863 (1967).
10. JACQUEZ, J. A., SCHAFER, J. A.: Biochim. Biophys. Acta 193, 368 (1969).
11. RIGGS, T. R., WALKER, L. M., CHRISTENSEN, H. N.: J. Biol. Chem. 233, 1479 (1958).
12. POTASHNER, S. J., JOHNSTONE, R. M.: Biochim. Biophys. Acta 203, 445 (1970).
13. POTASHNER, S. J., JOHNSTONE, R. M.: Biochim. Biophys. Acta 233, 91 (1971).
14. KUN TSAN LIN, JOHNSTONE, R. M.: Biochim. Biophys. Acta, in press (B. B. A.).
15. OXENDER, D. L., CHRISTENSEN, H. N.: J. Biol. Chem. 238, 3686 (1963).
16. BEGIN, N., SCHOLEFIELD, P. G.: Biochim. Biophys. Acta 90, 82 (1964).
17. BEGIN, N., SCHOLEFIELD, P. G.: J. Biol. Chem. 240, 332 (1965).
18. TALLAN, H. H., MOORE, S., STEIN, W. H.: J. Biol. Chem. 211, 927 (1954).

<u>Acknowledgements</u>

This work was supported by grants from the Medical Research Council of Canada. Much of the experimental work was done by Messrs. S. J. Potasher (Ehrlich cells) and Kun Tsan Lin (pancreas) in partial fulfilment of the Ph. D. degree, McGill University.

TABLE 1

THE $Na^+$ plus $K^+$ GRADIENT-INDUCED FLUX OF METHIONINE, GLYCINE, LEUCINE AND THIOUREA IN ATP-DEPLETED EHRLICH CELLS

Cells were depleted of ATP by incubation in $10^{-4}$ M 2, 4-dinitrophenol at $37^O$ in media containing 150 mM $K^+$ or 145 mM $Na^+$. The radioactive compounds were added after 30 min and the preincubation continued until the intracellular radioactivity approached the steady state (about 30-40 min). The cells were then centrifuged and added to fresh incubation medium which contained $10^{-4}$ M 2, 4-dinitrophenol in addition to the ionic composions given in the table. $1-^{14}C$ Glycine, L- $1-^{14}C$ leucine, $^{14}C$ thiourea and L- Me-$^{14}C$ methionine were added to the incubation medium at the same concentration (3 mM) and specific activity (120 counts/min per mole) as used during the preincubation period.

| Extracellular ($\mu$equiv/ml) | | Cellular $Na^+$ ($\mu$equiv/ml cell water) | | Cellular $K^+$ ($\mu$equiv/ml cell water) | | Net uptake (2-30 min)[+] (nmoles/mg dry wt.) | | | |
|---|---|---|---|---|---|---|---|---|---|
| $Na^+$ | $K^+$ | $t_2$ | $t_{30}$ | $t_2$ | $t_{30}$ | Methionine | Glycine | Leucine | Thiourea |
| 145 | 8 | 30-60 | 90-135 | 105-165 | 30-85 | 5.0 | 8.8 | 2.9 | 2.5 |
| 145 | 8 | 170-210 | 160-200 | 10-15 | 10-15 | 1.6 | 3.2 | - | - |
| Nil | 153 | 10-25 | 5-15 | 165-185 | 150-165 | -2.1 | -0.8 | - | - |
| Nil | 153 | 90-135 | 50-110 | 80-95 | 85-120 | -6.9 | -9.8 | -4.1 | -0.7 |

[+]Minus signs denote efflux.

TABLE 2

THE EFFECT OF ATP ON THE UPTAKE OF GLYCINE, METHIONINE,
LEUCINE AND THIOUREA INTO EHRLICH CELLS WITH NORMAL $Na^+$ plus $K^+$
GRADIENTS

ATP-depleted cells, containing known concentrations of the radioactive compounds
and high $K^+$ were obtained. The cells were transferred by fresh, normal Krebs-
Ringer medium containing $10^{-4}M$ 2, 4-dinitrophenol and the radioactive amino
acid at the same concentration and specific activity as that found intracellularly
at the end of the preincubation period. Generally the concentration and specific
activities were 3 mM and 120 counts/min per nmole, respectively. The incucations
were carried out at $25^O$ and samples were taken at several intervals over a 30
min period. Cells with normal ATP levels were used for comparative purposes.
The latter cells were preincubated in normal Krebs-Ringer for the same period
of time as the ATP-depleted cells. The cells were transferred to fresh medium
and incubated with the same concentration and specific activity of amino acids as
that used with ATP-depleted cells. Samples were taken until the intracellular concentration of radioisotope reached that of the medium. This point was considered
$t_o$ and equivalent to the starting point with the ATP-depleted cells. Incubation was
at $25^O$ with samples taken at intervals over a 30-min period.

| Additions | Inward $Na^+$ gradient ($\mu$ equiv/ml) | Outward $K^+$ gradient ($\mu$ equiv/ml) | Extracellular 2, 4-dinitrophenol ($\mu$ moles/ml) | Net uptake between 2 and 5 min (nmoles/mg dry wt.) |
|---|---|---|---|---|
| $1 - ^{14}C$ Gly | 95-105 | 145-155 | Nil | 8.4 |
|  | 70-100 | 85-145 | 0.1 | 0.8 |
| L- Met-$^{14}C$ | 95-105 | 145-155 | Nil | 3.7 |
|  | 70-100 | 85-145 | 0.1 | 0.5 |
| L- $1$-$^{14}C$ Leucine | 95-105 | 145-155 | Nil | 1.4 |
|  | 70-100 | 84-145 | 0.1 | 0.3 |
| $^{14}C$ Thiourea | 95-105 | 145-155 | Nil | 0.3 |
|  | 70-100 | 85-145 | 0.1 | 0.5 |

TABLE 3

THE Na$^+$ GRADIENTS, ATP AND THE UPTAKE OF METHIONINE FROM NORMAL KREBS-RINGER MEDIUM IN EHRLICH CELLS

Cells were preincubated in medium shown for 60 min and subsequently transferred to normal Krebs-Ringer medium containing 145 mequiv/l Na$^+$ and 8 mequiv/l K$^+$, 2 mM L-Me-$^{14}$C methionine: specific activity 125 counts/min per mole. Hepes buffer, pH 7.4, was used. Incubation was for 30 min at 25$^\circ$. Samples were taken at intervals between 0 and 30 min. Data for the 30-min values are given. 2,4-dinitrophenol, where used, was at 10$^{-4}$M. To elevate the ATP levels after treatment with 2,4-dinitrophenol, 10 mM glucose was added. The data given are typical of those obtained in three separate experiments. In all experiments wet and dry weight measurements were performed and cell volume determined. $t_o$ and $t_{30}$ represent the readings at 0 time and 30 min respectively.

| Preincubation conditions | Medium Na$^+$ ($\mu$equiv/ml) | Cellular Na$^+$ ($\mu$equiv/ml cell water) | | Cellular ATP ($\mu$moles/ml cell water) | | L-Me$^{14}$C methionine (uptake$^+$) (nmoles/mg dry wt.) |
|---|---|---|---|---|---|---|
| | | $t_o$ | $t_{30}$ | $t_o$ | $t_{30}$ | $t_{30}$ |
| 37$^\circ$ Krebs-Ringer | 145 | 20–30 | 35–50 | 2.1 | 2.8 | 24.0 |
| 10$^\circ$ K$^+$-Ringer | 145 | 2–4 | 35–50 | 2.1 | 2.8 | 26.0 |
| 10$^\circ$ Krebs-Ringer | 145 | 170–180 | 140–160 | 2.3 | 3.1 | 25.0 |
| 37$^\circ$ Krebs-Ringer + 2,4-dinitrophenol | 145 | 155 | 110 | 1.6$^{++}$ | 1.9$^{++}$ | 25.0 |
| 37$^\circ$ Krebs-Ringer + 2,4-dinitrophenol | 145 | 155 | 145 | 0.3 | 0.4 | 12.0 |
| 37$^\circ$ Krebs-Ringer + 2,4-dinitrophenol | 145 | 155 | 150 | 0.1$^{+++}$ | 0.1$^{+++}$ | 8.2 |

+ In fresh cells, the cell water was estimated to be 3.65 $\pm$ 0.04 (S.D.) $\mu$l/mg dry wt. Therefore in unswollen cells a cellular concentration of 2 mM is equivalent to 7.4 nmoles/mg dry wt.

++ The incubation medium contained 10 mM glucose.

+++ The incubation medium contained 10$^{-4}$ M 2,4-dinitrophenol.

## TABLE 4

THE $Na^+$ GRADIENT, ATP AND THE UPTAKE OF GLYCINE FROM NORMAL KREBS-RINGER MEDIUM IN EHRLICH CELLS

The experimental conditions are as outlined in Table III. 2 mM $1-^{14}C$ glycine, specific activity 125 counts/min per nmole was used.

| Preincubation conditions | Medium $Na^+$ ($\mu$equiv/ml) | Cellular $Na^+$ ($\mu$equiv/ml cell water) | | Cellular ATP ($\mu$moles/ml cell water) | | $1-^{14}C$ glycine uptake[+] nmoles/mg dry wt. |
|---|---|---|---|---|---|---|
| | | $t_o$ | $t_{30}$ | $t_o$ | $t_{30}$ | $t_{60}$ |
| $37^o$ Krebs-Ringer | 145 | 20–30 | – | $3.5^{++}$ | $3.1^{++}$ | 82.1 |
| $37^o$ Krebs-Ringer + 2,4-dinitrophenol | 145 | 150–160 | – | $2.5^{++}$ | $2.7^{++}$ | 74.7 |
| $37^o$ Krebs-Ringer + 2,4-dinitrophenol | 145 | 150–160 | – | 0.2 | 0.4 | 23.0 |
| $37^o$ Krebs-Ringer + 2,4-dinitrophenol | 145 | 150–160 | 150–160 | $0.1^{+++}$ | $0.1^{+++}$ | 16.4 |

+ A cellular concentration of 2 $\mu$moles/ml cell water of glycine is equivalent to 7.4 nmoles/mg dry wt. in the unswollen cell

++ The incubation medium contains 10 mM glucose.

+++ The incubation medium contains $10^{-4}$M 2,4-dinitrophenol.

TABLE 5

GLYCINE UPTAKE AND CATION CONTENT IN MOUSE PANCREAS in vitro

Values in parentheses are the number of experimental values. Glycine concentration was 2 mM (100 counts/min per nmol).

| Conditions during 60 min incubation | | Gas phase | Cation content ($\mu$equiv/ml cell water) | | Total glycine uptake ($\mu$moles/ml cell water $\pm$ S. D. | Intracellular/ extracellular ration of glycine |
| Medium | Additions | | $\pm$ S. D. Na$^+$ | K$^+$ | | |
|---|---|---|---|---|---|---|
| Krebs-Ringer | Nil | O$_2$ | 126 $+$ 14 (10) | 128 $+$ 13 (10) | 22.5 $+$ 3.7 (10) | 11.3 |
| | Nil | N$_2$ | 194 $+$ 3 (3) | 69 $+$ 11 (7) | 4.9 $+$ 0.4 (7) | 2.5 |
| | Ca$^{2+}$ | O$_2$/CO$_2$ §§§ | 144; 146 (2) §§ | 125; 132 (2) §§ | 21.1; 23.1 (2) §§ | 11.1 |
| | Nil | O$_2$/CO$_2$ §§§ | 126 $+$ 18 (5) | 139 $+$ 6 (5) | 21.0; 22.7 (2) §§ | 11.0 |
| | Dinitrophenol | O$_2$ | 186 $+$ 16 (4) | 72 $+$ 9 (4) | 9.6 $+$ O.9 (10) | 4.8 |
| +No K$^+$ | Nil | O$_2$ | 207 $+$ 8 (4) | 59 $+$ 2 (4) | 8.4 $+$ 0.9 (4) | 4.2 |
| | Dinitrophenol | O$_2$ | - | - | 4.1 $+$ 0.1 (4) | 2.6 |
| ++No Na$^+$ | Nil | O$_2$ | 9 $+$ 1 (4) | 60 $+$ 2 (4) | 4.2 $+$ 0.8 (7) | 2.1 |
| | Dinitrophenol | O$_2$ | - | - | 4.2 $+$ 0.4 (6) | 2.1 |
| +++Krebs-Ringer | 40 min pretreat- ment in N$_2$ | O$_2$ | - | - | 22.5; 24.0 (2) §§ | 11.5 |
| Fresh, unincubated mouse pancreas | | | 13 $+$ 2 (4) | 208 $+$ 3 (4) | - | - |

+ K$^+$ replaced by choline chloride.
++ Na$^+$ replaced by choline chloride.
+++ The tissue was preincubated for 40 min under N$_2$ before incubation at 37$^0$ in O$_2$ to measure uptake.
§ Dinitrophenol when present, at 10$^{-4}$ M; Ca$^{2+}$ at 2 mM.
§§ Duplicate values only.
§§§ NaHCO$_3$ buffer was used. All others Tris buffers.

TABLE 6

THE K⁺ GRADIENT AND GLYCINE ACCUMULATION IN MOUSE PANCREAS

Choline chloride was used to maintain isotonicity. Glycine concentration was 2 mM at 100 counts/min per mole. The values given are means $\pm$ S.D. or differences from the mean if only two values. Numbers in parentheses are the numbers of experimental values. Temperature of incubation: I and III, $37^\circ$; Expt. II, $25^\circ$. Duration of incubation: Expt. I, 30 min; Expt. II, 15 min; Expt. III, 60 min.

| Expt. | Medium cations ($\mu$equiv/ml) | | Cellular cations ($\mu$equiv/ml cell water) | | $\mathrm{Log}\dfrac{[K_i^+]}{[K_o^+]}$ | Total glycine uptake ($\mu$moles/ml cell water) |
|---|---|---|---|---|---|---|
| | $Na^+$ | $K^+$ | $Na^+$ | $K^+$ | | |
| I | 48 | 7.8 | $58 \pm 5$ (2) | $100 \pm 2$ (2) | 1.10 | $10.0 \pm 0.2$ (2) |
| | 48 | 105.0 | $52 \pm 3$ (2) | $180 \pm 4$ (2) | 0.23 | $9.2 \pm 0.2$ (2) |
| II | 87 | 7.8 | $127 \pm 2$ (3) | $93 \pm 6$ (3) | 1.05 | $3.7 \pm 0.3$ (2) |
| | 87 | 65.0 | $111 \pm 4$ (3) | $165 \pm 3$ (3) | 0.40 | $3.8 \pm 0.3$ (2) |
| III | 145 | 7.8 | $126 \pm 10$ (10) | $128 \pm 13$ (10) | 1.20 (control) | $24.3 \pm 1.8$ (4) |
| | 145 | 7.8 | $158 \pm 10$ (4) | $90 \pm 9$ (4) | 1.05 (ouabain)+ | $8.0 \pm 0.6$ (5) |

+ In Expt. III, ouabain was present at 0.2 mM where indicated.

## TABLE 7

ACTION OF GLUTAMINE ON MAINTAINING GLYCINE TRANSPORT ACTIVITY
IN MOUSE PANCREAS

The pancreas was preincubated in Krebs Ringer medium in $O_2$ except where indicated otherwise for 120 min with the additions as outlined in the table. Dinitrophenol was used at a concentration of $10^{-4}M$ and glucose at 5 mM. After the preincubation, the gas phase was changed to $O_2$ where required, $1-^{14}C$ glycine,
2 mM, at 88.7 counts/min per nmole was added and the incubation continued for
15 min at $37^O$. Values in parentheses represent the number of experimental values.
The variations are the ranges from the mean value.

| Preincubation conditions | | | Net glycine transport $(A_c-A_f)$ ($\mu$moles/ml cell water) minus ($\mu$moles/ml medium) |
|---|---|---|---|
| Time (min) | Gas phase | Additions | |
| 120 | $O_2$ | Standard medium | $9.3 \pm 1.5$ (4) |
| | | Dinitrophenol | $1.1 \pm 0.2$ (4) |
| | | Dinitrophenol + glucose | $1.0 \pm 0$ (2) |
| | | Dinitrophenol + glutamine (1 mM) | $6.7 \pm 0.2$ (3) |
| | | Dinitrophenol + glutamine (0.5 mM) | 4.3 (1) |
| | | Dinitrophenol + glutamate (1 mM) | $2.4 \pm 2$ (2) |
| | | Dinitrophenol + citrate (1mM) | $1.0 \pm 0$ (2) |
| | | Dinitrophenol + pyruvate (0.5 mM) | 1.04 (1) |
| 120 | $N_2$ | Glutamine (1 mM) | $0.4 \pm 0.1$ (2) |

# An Examination of the Energetic Adequacy of the Ion Gradient Hypothesis for Nonelectrolyte Transport

James A. Schafer

Departments of Physiology and Biophysics, and Medicine, The University of Alabama Medical Center, Birmingham, Alabama 35233, USA

A portion of the work reported here was completed while the author was a fellow of the Jane Coffin Childs Memorial Fund for Medical Research at the Institut für vegetative Physiologie, Frankfurt a. M., German Federal Republic. The author is currently an Established Investigator of the American Heart Association

As can be appreciated from the breadth of the investigations reported in this symposium, the topic of coupling between electrolyte and nonelectrolyte flows has been an active area in transport research. One of the primary bases for this interest has been the possibility that certain conservative nonelectrolyte transport processes (i. e., certain transport processes which demonstrate net flux against an apparent chemical potential gradient) might derive the required energy through a coupling to cation flows rather than directly to cellular metabolism. It has long been accepted that the asymmetrical distribution of $Na^+$ and $K^+$ across the cell membrane results from the operation of active transport processes which are coupled directly to cellular metabolism. Therefore, these phenomena may be referred to as "primary" active transport processes (Stein, 1967). According to the ion gradient hypothesis, the conservative transport of certain nonelectrolytes is assumed to be driven exclusively by coupling to the fluxes of $Na^+$ and $K^+$ down their electrochemical potential gradients. The nonelectrolyte uptake may be referred to as a "secondary" active transport process, since it has no direct coupling to metabolic energy (Stein, 1967). This general hypothesis has been supported most consistently by a number of investigations of amino acid and sugar transport in a variety of cells and epithelia (reviewed in detail by Schultz and Curran, 1970). Although there is a considerable body of evidence that cation gradients contribute energy to nonelectrolyte transport in these systems, it is still unresolved whether such coupling can provide all of the necessary energy. For example, from our recent investigations (Jacquez and Schafer, 1969; Schafer and Heinz, 1971) it appears that the ion gradients cannot meet entirely the energy requirement of amino acid active transport in the Ehrlich mouse ascites tumor cell, but that these gradients do contribute significantly. Whether this conclusion will be reached for other such transport systems is not yet clear. Furthermore, as discussed below, it must be appreciated that the interpretation of all such experiments suffers from uncertainty in several crucial areas, the resolution of which may alter even these tentative conclusions.

This presentation will discuss the issue in regard to, first, some models proposed for energetic coupling to nonelectrolyte transport; second, the thermodynamic requirements of coupled flows; third, a brief analysis of the experimental data as they relate to these topics; and last, some comments on the difficulties of interpretation involved.

## Models for Coupling Energy to Nonelectrolyte Transport

Before the ion gradient hypothesis had been formulated, it was assumed that all transport systems demonstrating net flux of a transported species against its electrochemical potential gradient must be coupled directly to cellular metabolism. (One exception was the counterflow phenomenon [Rosenberg and Wilbrandt, 1957] in which an apparent active transport resulted from preloading certain cells with another solute transported by the same system as the solute in question. Since counterflow could not give rise to a steady-state concentration gradient of either solute, it was easily distinguished from truly active transport processes ). Most investigators attributed active transport to a mobile carrier system, in which one of the reactions was made essentially irreversible by coupling it to the hydrolysis of ATP. One such scheme is illustrated in Fig. 1. Here it is assumed that the conversation of the carrier to a nontransporting form at the inside of the membrane, with the concomitant hydrolysis of ATP to ADP, constrains the carrier system to bring about a conservative uptake of the solute.

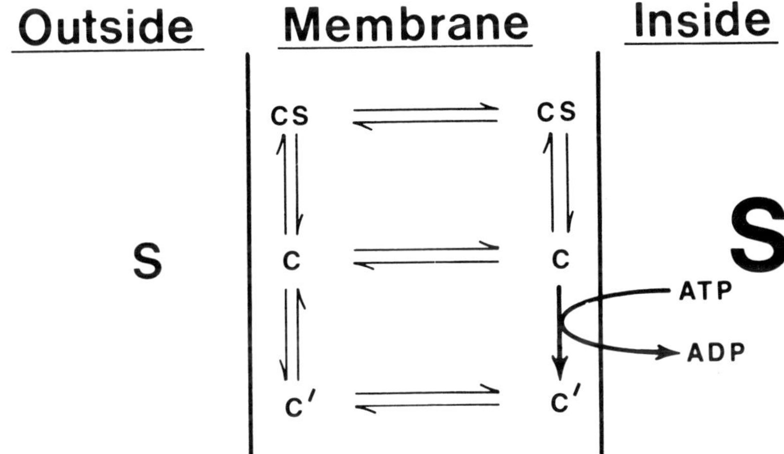


Fig. 1. A model for the coupling of cellular metabolism to the uptake of a nonelectrolyte. C represents a mobile carrier which has a site specific to certain nonelectrolytes designated by S. It is presumed that the carrier can be converted to a non-transporting form, C', with the concomitant hydrolysis of ATP at the inner membrane surface. The inactive carrier can only be reconverted to its transporting form through the action of a specific enzyme at the outer membrane surface. Because of the introduction of this irreversible step, the carrier cycle is constrained to bring about a unidirectional movement of the nonelectrolyte into the cell

An alternative view, the ion gradient hypothesis, was first proposed by Riggs, Walker and Christensen (1958) on the basis of several observations. As early as 1952, Christensen et al. (1952 a, b) observed that the uptake of certain amino acids in Ehrlich ascites cells and avian erythrocytes required $Na^+$ and was inhibited by $K^+$ in the extracellular medium, and that this uptake resulted in a gain of $Na^+$ and loss of $K^+$ by the cells. Later, Riggs et al. (1958) showed that the accumulation of glycine in these cells was dependent upon the relative intracellular concentrations of $Na^+$ and $K^+$ as well as their extracellular concentrations. In view of these findings, Riggs et al. (1958) proposed that glycine uptake might be coupled to a counterflow process in which $K^+$ exited from the cell via the glycine carrier, and by virtue of its asymmetrical distribution across the cell membrane, drove the inward transport of glycine (Fig. 2). They also suggested the alternative possibility that the process could be driven by a co-transport of $Na^+$ into the cell (Fig. 2). However, they preferred the former model partly on the basis that stimulators of amino acid uptake in these cells also accelerated $K^+$ efflux.

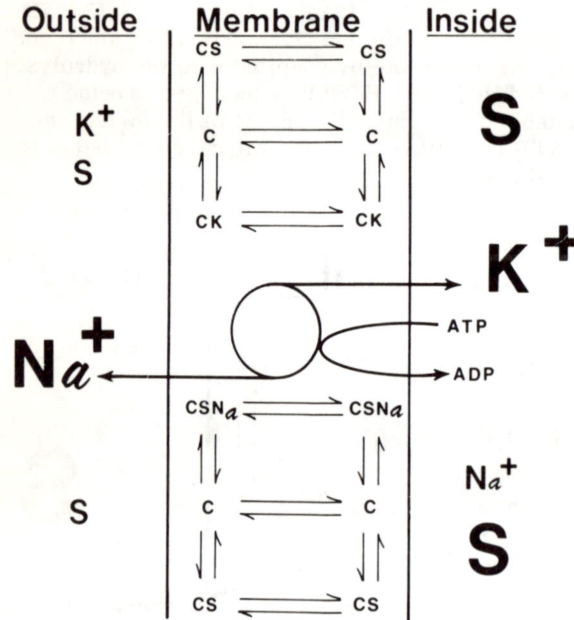


Fig. 2. Two models for the coupling of $Na^+$ or $K^+$ fluxes to nonelectrolyte uptake. In the upper part of the figure, the carrier C, is presumed to have a site capable of combining with the nonelectrolyte, S, or $K^+$. The nonelectrolyte will be concentrated in the cell by countertransport for $K^+$, which energetically drives the system by flowing down its electrochemical potential gradient. In the lower part of the figure, the carrier is assumed to form a complex with both $Na^+$ and the nonelectrolyte at the outer membrane surface. This complex dissociates at the inner membrane surface with the diffusion of free carrier back to the outer membrane surface. In this case, the nonelectrolyte is concentrated in the cell by the cotransport of $Na^+$ down its electrochemical potential gradient. In either case, the necessary energy is ultimately derived from the Na - K pump (cycle in center of figure) which utilizes metabolic energy by the hydrolysis of ATP. (Adapted from Riggs et al., 1958)

The $K^+$ gradient hypothesis later proved to be untenable on the basis of two observations. First, Hempling and Hare (1961) measured steady-state unidirectional fluxes of glycine and $K^+$ in Ehrlich cells, and showed that the energetic requirement for the maintenance of the glycine gradient exceeded the rate at which energy could be supplied by the dissipation of the $K^+$ electrochemical potential gradient. Secondly, Kromphardt et al. (1963) demonstrated that the influx of glycine was dependent primarily on the extracellular $Na^+$ rather than the extracellular $K^+$ concentration. This dependence of nonelectrolyte accumulation on $Na^+$ has been confirmed in almost all ion-dependent nonelectrolyte transport processes (Schultz and Curran, 1970).

In spite of the apparent primacy of $Na^+$ in such transport systems, a definite effect of $K^+$ can be demonstrated. In particular, replacement of extracellular $Na^+$ with $K^+$

inhibits nonelectrolyte transport to a greater degree than replacement of $Na^+$ with choline or Tris (Csaky, 1966; Kinis and Parrish, 1965; see also below). Crane (1962, 1965) attempted to reconcile these findings on the basis of a model in which both $Na^+$ and $K^+$ were involved in the nonelectrolyte transport process. He suggested that a carrier might have two sites: one for the nonelectrolyte, the other capable of accommodating either $K^+$ or $Na^+$. Furthermore, Crane proposed that when $Na^+$ occupied the ion site, the carrier had a high affinity for the nonelectrolyte, whereas with $K^+$ on the site the carrier had a low affinity for the nonelectrolyte. The model also involved the co-transport of $Na^+$ inward with the nonelectrolyte. Thus, nonelectrolyte transport could be coupled energetically to the $Na^+$ electrochemical potential gradient. However, this transport could not be coupled energetically to the $K^+$ gradient unless $K^+$ also fluxed out of the cell down its electrochemical potential gradient via the carrier. This possibility was not proposed in the original Crane hypothesis.

A consideration of this model has suggested to some investigators that both ion gradients might contribute to nonelectrolyte transport especially in view of a number of findings implicating $K^+$ in this system:

1. There is considerable evidence that the net accumulation of several nonelectrolytes is depressed either by removing $K^+$ from the extracellular medium or by raising its concentration above 4 mEq liter$^{-1}$ (Christensen, 1952a; Riklis and Quastel, 1958; Kleinzeller and Kotyk, 1961; Eddy, 1968a, b).

2. There is an apparent competition between $Na^+$ and $K^+$ for the carrier site (Kipnis and Parrish, 1965; Eddy and Hogg, 1969; Crane, 1965).

3. Amino acid uptake in the Ehrlich ascites cell increases both $K^+$ influx and efflux (Christensen, 1952b; Hempling and Hare, 1961; Eddy, 1968a, b), although this admittedly could be secondary to the primary $Na^+$ movements (Schultz and Curran, 1970; Jacquez, 1971).

4. Transient steady-state gradients of alpha-aminoisobutyric acid (AIB) and glycine seem to be correlated better with the sum of the $Na^+$ and $K^+$ electrochemical potential gradients than with the $Na^+$ gradient alone (Eddy and Hogg, 1969; Schafer and Heinz, 1971).

Such implications of $K^+$ involvement derive mainly from studies of amino acid transport in Ehrlich cells and striated muscle, and sugar transport in the gut; there seems to be no important $K^+$ effect for amino acid transport in the gut (Schultz and Curran, 1970), pigeon red cells (Vidaver, 1964a, b, c, d), or rabbit reticulocytes (Wheeler and Christensen, 1967). As pointed out by Eddy and Hogg (1969), the additional coupling of the $K^+$ gradient would enable the systems to concentrate the nonelectrolyte to a greater extent than with $Na^+$ gradient coupling alone. Thus, although there is no conclusive evidence for the coupling of the $K^+$ gradient, we should not neglect its possible contribution to the energetic requirements of any conservative nonelectrolyte transport process. The model shown in Fig. 3 illustrates a coupling of nonelectrolyte transport to both the $Na^+$ and $K^+$ electrochemical potential gradients; however, this model is not meant to define the precise mechanism of kinetics of such a coupling, but rather to provide a conceptual framework in which to discuss the energetic consequences of the parallel movements of $Na^+$ and the nonelectrolyte, and the oppositely directed movement of $K^+$.

## The Thermodynamic Requirements of a Coupled Flow vs an Active Transport System

Until the possibility of coupled solute flows was considered, a sufficient criterion for an active transport process was that it brought about net movement of a transported species against its electrochemical  potential gradient. However, such a criterion requires that water transport against an osmotic gradient be considered an active transport process, whereas it is well established that the movement of water against a gradient results from the coupling of water flow to active $Na^+$ transport, e. g. the transport of water against an osmotic gradient in the gallbladder (Durbin et al., 1956). Since we cannot regard water transport as active  in such circumstances, the conservative uptake of a nonelectrolyte which is driven completely by coupling to cation flows cannot be termed active (Heinz,  1960 and 1963). Kedem (1961) has formalized this requirement in the phenomenologic relation:

$$J_i = \frac{-\Delta\widetilde{\mu}_i}{R_{ii}} + \sum_{\substack{k=1 \\ k \neq i}}^{n} \frac{R_{ik}}{R_{ii}} J_k + \frac{R_{ir}}{R_{ii}} J_r , \qquad (1)$$

where: $J_i$ is the flux of the solute in question, $\Delta\widetilde{\mu}_i$ is the electrochemical potential difference of this solute across the membrane, $J_k$ is the flux of any other solute or water, $J_r$ is the rate of a metabolic reaction which may be coupled to $J_i$, and $R_{ik}$ and $R_{ii}$ represent the coefficients of resistance relating the i-th force to the k-th or i-th flow, respectively. For any other flux to which $J_i$ is coupled, the coefficient $R_{ik}$ will be non-zero. If $J_i$ is also coupled to cellular metabolism, $J_r$ and $R_{ir}$ will also be non-zero. Kedem (1961) has defined an active transport system as one in which these latter conditions obtain. In her terminology, a system could show a flux of solute, $J_i$, against an electrochemical potential gradient if the solute flux were coupled appropriately to some other flow, or if it were coupled to cellular metabolism; only in the latter case is the transport considered active.

If we apply Kedem's requirement to a nonelectrolyte transport which shows cation dependence, we have several possible flows to which the transport might be linked. It could be linked to the inward flux of $Na^+$, the outward flux of $K^+$, or, by the phenomenon of counterflow, to the net flux of similar solutes traversing the membrane via the same carrier. Lastly, there is the possibility that the transport could be linked additionally to cellular metabolism. Thus, the net  flux of the given nonelectrolyte, $J_i$, would be given by the relation:

$$J_i = -\frac{\Delta\widetilde{\mu}_i}{R_{ii}} + \frac{R_{iN}}{R_{ii}} J_N + \frac{R_{iK}}{R_{ii}} J_K + \sum_{\substack{j=1 \\ j \neq i}}^{n} \frac{R_{ij}}{R_{ii}} J_j + \frac{R_{ir}}{R_{ii}} J_r, \qquad (2)$$

where: $J_N$ is the coupled flux of $Na^+$, $J_K$ is the coupled flux of $K^+$ and $J_j$ is the flux of any solute exchanging via the same carrier, and $R_{iN}$, $R_{iK}$ and $R_{ij}$ are the respective resistances.[1] If the system is coupled only to a cation flow, it must be shown that either or both $R_{iN}$ and $R_{iK}$ are non-zero, and that $R_{ir}$ is zero. The

---

[1] It should be noted that the resistance coefficients will take signs appropriate to the mechanism by which the coupling to the nonelectrolyte flux is affected. For example, if $Na^+$ is co-transported while $K^+$ is countertransported with respect to the nonelectrolyte (as in Fig. 3), $R_{iN}$ and $R_{iK}$ would have opposite signs.

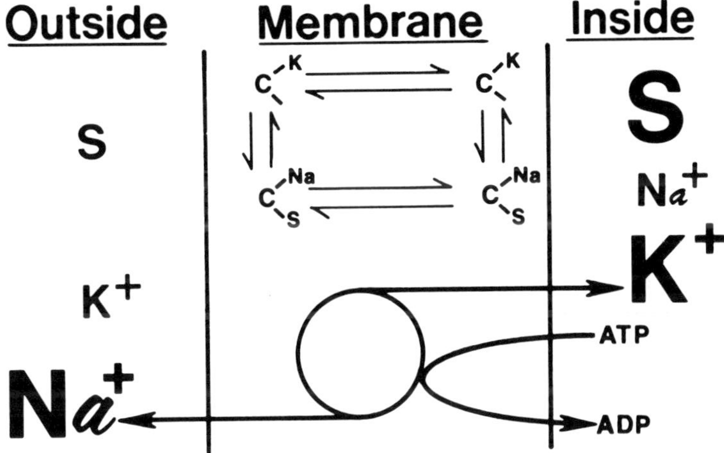



Fig. 3. A modification of the Crane hypothesis for the coupling of nonelectrolyte transport to $Na^+$ and $K^+$ electrochemical potential gradients. The model as presented by Crane (1965) hypothesized a two-site carrier, C, one site of which combined with the nonelectrolyte, S, the other site accommodating $Na^+$ or $K^+$. At the outer membrane surface, where the $Na^+$ concentration was high, the $Na^+$-nonelectrolyte complex predominated and diffused to the inner membrane surface. Here, in the presence of a high $K^+$ concentration, $K^+$ replaced $Na^+$ on the ion site. It was further hypothesized that with $K^+$ bound to the ion site, the carrier had a much lower nonelectrolyte affinity, which resulted in the dissociation of the nonelectrolyte. The additional possibility that $K^+$ could flux out of the cell in combination with the carrier has been added to this original Crane model. In this way, the concentrative uptake of the nonelectrolyte can be driven by cotransport of $Na^+$ and countertransport of $K^+$, both down their respective concentration gradients. The ultimate energy source is the utilization of cellular metabolism through the hydrolysis of ATP by the Na - K pump, as shown in the lower part of the figure

further possibility that all the $J_j$' s and/or $R_{ij}$' s are not zero complicates the analysis. However, equation 2 does make one important point quite obvious. In order for the nonelectrolyte flux to be coupled energetically to either the $Na^+$ or $K^+$ electrochemical potential gradient, there must be movements of the two ions across the membrane concomitant with the nonelectrolyte movement. This is understandable from a thermodynamic standpoint since a potential energy gradient cannot, by its mere existence, provide energy. Rather, if the energy inherent in this gradient is to be utilized, it must be dissipated.

Thus, according to the ion gradient hypothesis, the conservative uptake of a nonelectrolyte is driven energetically by the fluxes of $Na^+$ and $K^+$ down their respective electrochemical potential gradients, and the consequent dissipation of the energy inherent in these gradients. If one considers the Crane hypothesis in these terms, it is obvious that the $K^+$ electrochemical potential gradient will contribute energetically to the sugar transport only if $K^+$ actually moves out of the cell on the carrier. Although the high intracellular $K^+$ concentration might act to displace $Na^+$ and sugar from the carrier, it would not couple any additional energy to the process. Rather, this effect of $K^+$ would serve only to tighten the coupling between the sugar uptake and the dissipation of the $Na^+$ gradient, by reducing the back flux of carrier complexed with sugar, or carrier complexed with sugar and $Na^+$.

Another aspect of these thermodynamic considerations concerns one of the more seductive arguments in favor of the ion gradient hypothesis, which contends that this mechanism is energetically more economical for the cell than the direct coupling of cellular metabolism to both cation and nonelectrolyte transport. Equation (2) points out clearly the fallacy in this argument. Since the coupling of ion fluxes to the nonelectrolyte flux is a real process, it must be irreversible, i. e. , not all of the energy dissipated by the $Na^+$ and $K^+$ fluxes can be utilized to drive the nonelectrolyte transport. Therefore, the final efficiency of utilization of ATP energy by the nonelectrolyte transport process could be considerably less than if ATP hydrolysis were coupled directly to this transport with an equivalent reduction in the dissipative leaks of $Na^+$ and $K^+$ through the nonelectrolyte system. In this case, if any economy were realized by the cell, it would be that the enzymatic "machinery" for the coupling of ATP hydrolysis would be required for only one transport process.

Returning to the phenomenological description of coupled transport, we can also describe the influx of a nonelectrolyte in terms of the forces to which it is linked, that is, its own electrochemical potential gradient, the electrochemical potential gradients of $Na^+$ ($\Delta \tilde{\mu}_N$) and $K^+$ ($\Delta \tilde{\mu}_K$), and any exchanging solutes ($\Delta \tilde{\mu}_j$), and possibly the affinity of the metabolic reaction to which it may be linked (A; Katchalsky and Curran, 1965):

$$J_i = - L_{ii} \Delta \tilde{\mu}_i - L_{iN} \Delta \tilde{\mu}_N - L_{iK} \Delta \tilde{\mu}_K - \sum_{\substack{j=1 \\ j \neq i}}^{n} L_{ij} \Delta \tilde{\mu}_j + L_{ir}A, \quad (3)$$

where the L's represent the coupling coefficients between $J_i$ and the various forces.

It should be noted that equation (3) utilizes the electrochemical potential gradients of $Na^+$ and $K^+$. This implies that the membrane potential may also influence the transport of the nonelectrolyte. The evidence on this point is scant. Vidaver (1964c) has shown that, in hemolyzed and restored pigeon red blood cells with equal intra- and extracellular $Na^+$ concentrations, the imposition of a Donnan potential across the membrane by means of adding the large anion mucate to the extracellular medium results in a concentrative extrusion of glycine from the cells. In view of these data, it seems prudent to consider the possibility that the membrane potential may indeed contribute to the process.

Three of the main means of examining the energetic adequacy of the ion gradient hypothesis are best understood in terms of equation (3):

i) <u>The steady-state test</u> - Under steady-state conditions, all non-actively transported solutes will be expected to be distributed with $\Delta \tilde{\mu}_j$ - 0, and there will be no net solute flux ($J_i = 0$). Under such conditions, if the ion gradients are the sole energy source (i. e. , $L_{ir} = 0$), the gradient of the solute will be directly proportional to the electrochemical potential gradient of $Na^+$ and the oppositely directed electrochemical potential gradient of $K^+$, and will be zero if their sum is zero.

ii) <u>Metabolic inhibition</u> - The same will apply with steady-state analyses under conditions of metabolic inhibition. However, if one assumes that the given means of metabolic inhibition abolishes all the high energy substrates for the transport system, then the $L_{ir}$ term would be zero.

iii) <u>Uptake studies with reversed ion gradients</u> - When the $Na^+$ and $K^+$ gradients are reversed, then both would tend to oppose the conservative uptake of the nonelectrolyte. If, under these conditions, conservative uptake occurs, it must be due either to the contribution of counterflow or to coupling to metabolism.

The next section deals with the application of these methods to assess the energetic adequacy of the ion gradient hypothesis in several ion-dependent transport systems.


## Experimental studies

Relatively little work has been done on the problem of the energetic adequacy of the cation gradients for nonelectrolyte active transport, and evidence bearing directly on this problem is found for relatively few $Na^+$ dependent transport processes. Those for which such data do exist include amino acid transport in avian erythrocytes, Ehrlich ascites cells and the intestine, and sugar transport in the intestine.

Vidaver (1964a, b, c, d) has presented compelling evidence for the energetic adequacy of the $Na^+$ gradient for a nonelectrolyte transport process. Working with glycine transport in hemolyzed and restored pigeon red blood cells, he established that the magnitude of the quasi-steady-state glycine concentration gradient developed in these cells was proportional to the magnitude and direction of $Na^+$ ion gradient. When $Na^+$ was distributed with equal intra- and extracellular concentrations, glycine was distributed identically, but, as noted above, the imposition of a Donnan membrane potential under these circumstances gave rise to a net glycine efflux. Since ATP levels in these cells were extremely low, the quasi-steady-state distribution of glycine away from chemical equilibrium had to be attributed to the $Na^+$ electrochemical potential gradient.

Similar studies were conducted by Eddy et al. (1967) and Eddy (1968a, b) on Ehrlich ascites cells. In his studies Eddy depleted cells of ATP by incubating them with NaCN, and, by varying the initial incubation procedures, he was able to control the intracellular cation levels. He then measured the "transient steady-state" glycine distribution reached in various extracellular media, and found that there was a direct dependence of the glycine distribution on the $Na^+$ gradient across the membrane. Thus, in spite of the absence of ATP, glycine could be concentrated in the cells as long as the $Na^+$ gradient was directed inwardly. However, because the glycine distribution ratio generally exceeded the $Na^+$ distribution ratio, Eddy (1968b, 1969) hypothesized an additional energetic contribution from the $K^+$ ion gradient, although it was significantly smaller than that of the $Na^+$ gradient.

The experiments of both Eddy and Vidaver argue strongly for at least a contribution of the $Na^+$, or both the $Na^+$ and $K^+$ gradients, to the uptake of amino acids in these systems. However, it must be recognized that both experimental methods employed cells depleted of ATP. Consequently, no coupling to cellular metabolism could be expected, and a sole dependence of the uptake on the ion gradients could not be demonstrated conclusively. An important point bearing on this question is Eddy's finding (1968b) that in NaCN-inhibited cells under optimal cation gradient conditions, the steady-state glycine distribution reached was only 30% of that seen in normal respiring cells with similar gradients. One possible explanation for this effect (Eddy, 1968b) is that in respiring cells the amino acid transport is coupled more tightly to the $K^+$ gradient; however, this explanation is no more satisfactory than the possibility that cellular metabolism contributes to the uptake in normally respiring cells.

Because of the technical complexities, similar experiments in the intestine have been more restricted. Crane (1964) incubated hamster small intestine fragments in media containing 6-deoxyglucose and 2, 4-dinitrocresol in the absence of oxygen. After these cells had accumulated relatively high $Na^+$ concentrations and equilibrated with the 6-deoxyglucose, they were resuspended in a $Na^+$ free medium containing the same 6-deoxyglucose concentration. Under these conditions, with a strongly reversed gradient of $Na^+$, net sugar efflux against a concentration gradient was demonstrated. However, due the presence of the metabolic inhibitor, any contribution from cellular metabolism was precluded.

More recently Hajjar et al. (1970) have shown that mucosal strips from rabbit ileum treated with ouabain or NaCN demonstrate net alanine transport which is dependent upon the direction and magnitude of the $Na^+$ gradient. These investigators felt, justifiably, that their results definitely were consistent with the hypothesis that the $Na^+$ gradient provides at least part of the energy for alanine uptake in this system, but there was insufficient evidence to conclude that it was the sole energy source.

More recent studies by Kimmich (1970a, b) have added additional complexities to the application of the ion gradient hypothesis to amino acid transport in the intestine. In his experiments, Kimmich employed isolated mucosal cells from chicken gut, and was able to demonstrate a high degree of conservative uptake in the presence of reversed ion gradients. Furthermore, he could demonstrate relatively little effect of the orientation of these gradients on the uptake processes. These findings might be rationalized on the basis of species differences, although a more appealing solution should be sought.

Similar results to those of Kimmich - but not as extreme - have been reported for amino acid transport in other cell suspensions. Wheeler and Christensen (1967) reported a concentrative uptake of alanine in rabbit reticulocytes in spite of an outwardly directed $Na^+$ gradient. We (Schafer and Jacquez, 1967, 1968; Jacquez and Schafer, 1969) have also reported experiments in which either the $Na^+$ alone or both the $Na^+$ and $K^+$ electrochemical potential gradients were reversed in Ehrlich ascites cells through the use of ouabain or cold incubation procedures; however, in spite of these reversals, conservative uptake of alpha-aminoisobutyric acid (AIB) could be demonstrated. Although these results appear to contradict directly the ion gradient hypothesis, another possible driving force other than metabolism may be involved. The Ehrlich ascites cell is known to have large pools of endogenous amino acids (Oxender, 1965) which might drive AIB influx by counterflow. This would be equivalent to a coupling between $J_i$ and $J_j$ in equation (2). As discussed by Jacquez (1971), we attempted to circumvent this possibility by examining the steady-state relationship of the AIB, $Na^+$ and $K^+$ electrochemical potential gradients (Jacquez and Schafer, 1969). This analysis demonstrated that the energy available in the $Na^+$ electrochemical potential gradient fell far short of satisfying the energy requirements for the maintenance of the steady-state AIB gradient. The combined $Na^+$ and $K^+$ electrochemical potential gradients came much closer to this energy requirement. However, it should be stressed that these data demonstrated that in order for the combination of the $Na^+$ and $K^+$ gradients to satisfy the entire energy requirement, they had to be coupled to the amino acid transport system with 100% efficiency, a demand which seems rather improbable.

Potashner and Johnstone (1970, 1971) have reported additional evidence bearing on the energetic coupling of amino acid transport in the Ehrlich ascites cell. They demonstrated that although glycine and methionine transport in these cells was dependent upon the magnitude and direction of the ion gradients in the presence of very low cellular ATP levels, the restoration of normal ATP levels gave a much greater amino acid uptake. Thus, ATP seemed to be a more important determinant of amino acid transport than the ion gradients, and was apparently the larger contributor to the required energy.

Recently, we reported experiments involving the transport of AIB in the Ehrlich ascites cell (Schafer and Heinz, 1971). In these experiments, the cells were incubated for 30 minutes with 2 mM AIB (labeled with $^{14}$C), resulting in a near steady-state distribution. The cells were exposed to a series of incubations in full isotonic Krebs Ringer phosphate medium at 0°C , alternated with incubations in one-half isotonic Krebs Ringer phosphate medium at 0°C. In some experiments, the media also contained 0.5 mM ouabain, but in all cases the procedures resulted in higher cell $Na^+$ concentrations, with lower cell $K^+$ and AIB concentrations. The cells were then incubated for 3 minutes at 37°C. in media of varying ionic composition, which contained, in addition, AIB at the same specific activity as in the initial incubation. Thus, this final incubation could be carried out under a variety of electrochemical potential gradients of $Na^+$, $K^+$, and AIB itself. Table I shows some representative results of such experiments. This data demonstrated clearly that the observed uptake of AIB correlated much more closely with the sum of $Na^+$ and $K^+$ electrochemical potential gradients than with the $Na^+$ gradient alone. However, even when both gradients were reversed, a net conservative uptake of AIB was observed. When ouabain was employed, the gradients were much more strongly reversed, and in this situation a net efflux of AIB against its concentration gradient could be demonstrated.

In order to evaluate thermodynamically the implications of these data, we (Schafer and Heinz, 1971) proposed the following relationship based on equation (3):

$$J_A = \nu_A{}^2 \, L_{rr} \, (\Sigma X_{vec} + A_r / \nu_A), \qquad (4)$$

where: $J_A$ is the AIB flux, $\nu_A$ is the constant stoichiometric coefficient for AIB in the transport reaction, $A_r$ is the affinity of a chemical reaction (such as the hydrolysis of ATP) which might be associated with the transport process; and $\Sigma X_{vec}$ is the sum of the vectorial forces tending to concentrate AIB, that is, the electrochemical potential gradient of $Na^+$ and AIB, and the oppositely-directed electrochemical potential gradient of $K^+$. If one plots $J_A$ vs. $\Sigma X_{vec}$, such a plot will go through the origin if $A_r = 0$, that is, if there is no driving force in addition to the vectorial driving forces. The actual plot of all the data without ouabain is shown in Fig. 4, which shows that there is a highly significant X and Y intercept. Stated in another way, there is substantial net transport of AIB when the sum of the vectorial driving forces is zero, and this uptake does not become zero until the sum of the net driving forces is greater than 4000 joules mole$^{-1}$ in opposition to the net transport. From this analysis, it appears that an additional driving force, of the order of 4000 joules per mole, is required to explain the conservative uptake of AIB under these gradient conditions. In the presence of 0.5 mM ouabain, a similar plot is obtained but with an energy deficiency of only 2200 joules mole$^{-1}$. The most likely source of this energy deficit appears to be coupling to cellular metabolism. If this were so, one must reconcile the additive contributions of the ion electrochemical potential gradients and the coupling to cellular metabolism. One model for such a combined energy supply is shown in Fig. 5. In this model, the ion gradients drive the system in the same way as depicted in Fig. 3; however, a parallel pathway involving an irreversible conversion of the carrier with the hydrolysis of ATP, which is similar to the model of Fig. 1, is added. Under conditions of metabolic inhibition, such as used by Eddy (1968), the movement of amino acid would be responsive en-

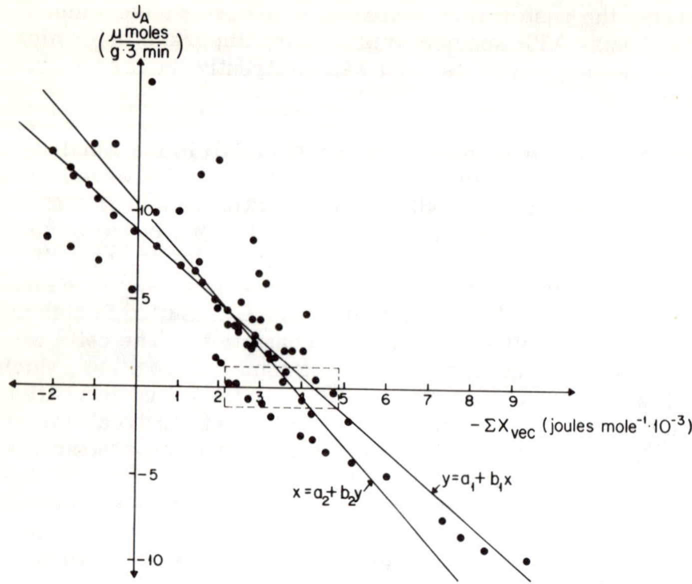



Fig. 4. The net flux of AIB as a function of the sum of the vectorial driving forces. $J_A$ is the net flux of AIB in $\mu$ moles per 8 g dry weight over a 3-minute period. $J_A$ is positive for a net influx. $\sum X_{vec}$ is the sum of the vectorial driving forces taken as the AIB chemical potential gradient, the $Na^+$ electrochemical potential gradient, and the negative of the $K^+$ electrochemical potential gradient. The two lines represent the least squares regression of y on x ($r = 0.88$) and x on y ($r = 0.88$). The dotted lines enclose points for which $J_A$ is approximately zero. (Taken from Schafer and Heinz, 1971, with permission of the publisher.)

tirely to the ion gradients. On the other hand, in normal cells, reversal of the ion gradients would lead to a cessation or reversal of uptake only if their opposing energetic contribution exceeded the energetic contribution of the metabolically coupled step. Certainly, other models might be equally valid in explaining this phenomenon; however, this model does combine some of the essential features of both the ion gradient and metabolically linked transport processes, and, if further assumptions were made about the influence of the species bound at the ion site on the affinity of the amino acid site, it could predict all of the observed kinetic phenomena.

It seems pointless to carry such speculation much further in view of the several limitations to the interpretation of data derived from all experiments in which the electrochemical potential gradients of $Na^+$, $K^+$ and a nonelectrolyte are determined. These limitations, discussed below, have been examined by Schafer and Heinz (1971) and Jacquez (1971).

First, there is the possibility that the uptake of the nonelectrolyte may be coupled to other flows than those considered. Steady-state experiments such as those of Jacquez and Schafer (1969) and experiments involving a preequilibration of the nonelectrolyte to a near steady-state (Schafer and Heinz, 1971) should exclude the energetic contribution of any exchanging solute which is not transported actively. However, the possibility still remains that protein hydrolysis or other metabolic activity might provide sufficient intracellular exchanging solutes to drive a conservative uptake of the nonelectrolyte.

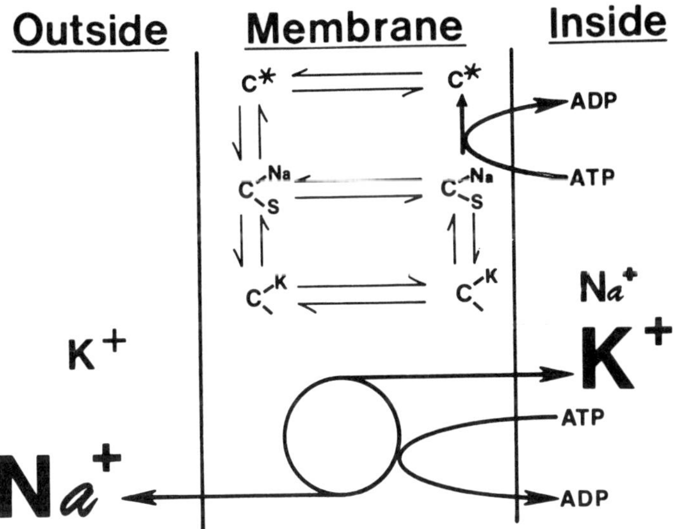


Fig. 5. A model for the coupling of nonelectrolyte transport to the electrochemical potential gradients of $Na^+$ and $K^+$, and to metabolism. As in the model of Fig. 2, a two-site carrier, C, which can accommodate $Na^+$ or $K^+$ at the ion site is presumed. However, in addition, it is presumed that the carrier can be converted to a nontransporting form, $C^*$, through the hydrolysis of ATP in Fig. 1. Consequently, the accumulation of S by the cell is driven by the fluxes of $Na^+$ and $K^+$ down their electrochemical potential gradients and the direct coupling to cellular metabolism

Second, the most critical problem involves our ability to determine the $Na^+$, $K^+$ and nonelectrolyte activities at the intracellular membrane surface. The calculations employed in every laboratory assume that the total solute analyzed is distributed evenly in the cell water and freely soluble. In fact, Cope (1970) has presented NMR evidence that a substantial portion of the intracellular $Na^+$ in mammalian muscle, brain and kidney is bound. Also Siebert et al. (1965) and Pietrzyk and Heinz (1972) report that nuclear $Na^+$ concentrations are considerably higher than cytoplasmic $Na^+$ concentrations in liver and Ehrlich ascites cells, at least in media of normal cation composition. Such distributions or binding of the intracellular $Na^+$, $K^+$ or nonelectrolyte could alter greatly the interpretation of experiments with reversed ion gradients (Jacquez, 1971; Schafer and Heinz, 1971). On the other hand, recent careful experiments by Krolenko and Adamjan (1967) have demonstrated that single muscle fibers behave osmotically as if their intracellular electrolytes had the same activity at the cell membrane as would be computed if the total amount of each ion were evenly distributed over the entire cell water. Also, Hinke (1961) has measured intracellular $Na^+$ and $K^+$ activities in squid axon with microelectrodes and has arrived at the same conclusion.

Even if such evidence cannot be applied to the Ehrlich ascites cell, Jacquez (1971) has calculated that the actual cytoplasmic $Na^+$ activity would have to be 17 to 33 % of that calculated in order for the $Na^+$ gradient to just satisfy the energetic requirements of AIB steady-state distributions observed in this cell. Furthermore, the fact that Eddy (1968a, b) and Vidaver (1964a, b, c, d) obtain a direct correspondence between the cation and amino acid gradients without any metabolic contribution, suggests either that metabolism is required for the uneven distribution or binding, or that they do not exist.

Third, in experiments such as those reported by Jacquez and Schafer (1969) and Schafer and Heinz (1971), it is assumed tacitly that the contribution from metabolism, if present, will be the same, regardless of the various cation gradients to which the cell is exposed. Actually, Heald (1954) has shown that high extracellular $K^+$ concentrations uncouple oxidative phosphorylation, and Abadom and Scholefield (1961) have shown that, under high extracellular $K^+$ concentrations, there is a reduction of glycine uptake by rat brain cortex coincident with a reduction in cellular ATP levels. Thus, our experiments in which both ion gradients are reversed may also occasion decreased cellular ATP levels with a subsequent reduction in amino acid uptake or an actual conservative amino acid efflux. If such were the case, the energy deficit reported by Schafer and Heinz (1971) might be even greater in the presence of normal extracellular media. In spite of these limitations, it appears that even the most severe errors in the estimation of the $Na^+$ gradient could not account for the failure of this gradient to explain the entire energetic demands of AIB transport in the Ehrlich cell. Further, in order for the $Na^+$ and $K^+$ gradients to be the sole energy source in this transport process, similar errors, of considerable magnitude, in the $K^+$ and AIB gradients would be required.

The most urgent step in the further investigation of this problem is, therefore, dependent upon a thorough investigation of ion distribution and activity in the cell. Until such data are forthcoming, it seems appropriate to suggest that, at least in the Ehrlich cell, the conservative uptake of amino acids cannot be driven completely by the $Na^+$, or a combination of the $Na^+$ and $K^+$, electrochemical potential gradients, and that some additional energy supply, most likely cellular metabolism, is involved.

## References


ABADOM, P.N., SCHOLEFIELD, P.G.: Amino acid transport in brain cortex slices. III. The utilization of energy for transport. Can. J. Biochem. Physiol. **40**, 1603-1618 (1962).

CHRISTENSEN, H.N., RIGGS, T.E., FISCHER, H., PALATINE, I.N.: Amino acid concentration by a free cell neoplasm: Relations among amino acids. J. Biol. Chem. **198**, 1-15 (1952a).

CHRISTENSEN, H.N., RIGGS, T.E., RAY, N.E.: Concentrative uptake of amino acids by erythrocytes in vitro. J. Biol. Chem. **194**, 41-51 (1952b).

COPE, F.W.: Spin-echo nuclear magnetic resonance evidence for complexing of sodium ions in muscle, brain and kidney. Biophys. J. **10**, 843-858 (1970).

CRANE, R.K.: Hypothesis for the mechanism of intestinal active transport of sugars. Fed. Proc. **21**, 891-895 (1962).

CRANE, R.K.: Uphill outflow of sugar from intestinal epithelial cells induced by reversal of the $Na^+$ gradient: Its significance for the mechanism of $Na^+$-dependent active transport. Biochem. Biophys. Res. Comm. **17**, 481-485 (1964).

CRANE, R.K.: $Na^+$-dependent transport in the intestine and other animal tissues. Fed. Proc. **24**, 1000-1006 (1965).

CSAKY, T.F.: Effect of mucosal potassium on the intestinal glucose transport. Pflügers Archiv **291**, 63-68 (1966).

DURBIN, R.P., FRANK, H., SOLOMON, A.K.: Water flow through frog gastric mucosa. J. Gen. Physiol. **39**, 535-551 (1956).

EDDY, A.A., MULCAHY, M.F., THOMSON, P.J.: The effects of sodium ions, potassium ions on glycine uptake by mouse ascites-tumor cells in the presence and absence of selected metabolic inhibitors. Biochem. J. **103**, 863-876 (1967).

EDDY, A.A.: A net gain of sodium ions and a net loss of potassium ions accompanying the uptake of glycine by mouse ascites-tumour cells in the presence of sodium cyanide. Biochem. J. 108, 195-206 (1968a).

EDDY, A.A.: The effects of varying cellular and extracellular concentrations of sodium and potassium ions on the uptake of glycine by mouse ascites tumour cells in the presence and absence of sodium cyanide. Biochem. J. 108, 489-498 (1968b).

EDDY, A.A., HOGG, M.C.: Further observations on the inhibitory effect of extracellular potassium ions on glycine uptake by mouse ascites-tumour cells. Biochem. J. 114, 807-814 (1969).

HAJJAR, J.J., LAMONT, A.S., CURRAN, P.F.: The sodium-alanine interaction in rabbit ileum. J. Gen. Physiol. 55, 277-296 (1970).

HEALD, P.J.: Rapid changes in creatine phosphate level in cerebral cortex slices. Biochem. J. 57, 673-679 (1954).

HEINZ, E.: Grundmechanismus der Magensäureproduktion und deren Regulation. Klin. Physiol. I, 184-203 (1960).

HEINZ, E.: Physiologie, Biochemie und Energetik des aktiven Transports. Naunyn-Schmiedeberg's Arch. exp. Path. Pharmak. 245, 10-28 (1963).

HEMPLING, H.G., HARE, D.: The effect of glycine transport on potassium flux in the Ehrlich mouse ascites tumor cell. J. Biol. Chem. 236, 2498-2502 (1961).

HINKE, J.A.M.: The measurements of sodium and potassium activities in the squid axon by means of cation selective glass microelectrodes. J. Physiol. 156, 314-335 (1961).

JACQUEZ, J.A., SCHAFER, J.A.: $Na^+$ and $K^+$ electrochemical potential gradients and the transport of $\alpha$-aminoisobutyric acid in Ehrlich ascites tumor cells. Biochim. Biophys. Acta 193, 368-383 (1969).

JACQUEZ, J.A.: Ion gradient hypotheses and the energy requirement for active transport of amino acids. In this symposium.

KEDEM, O.: Criteria of active transport. In "Membrane Transport and Metabolism", A. Kleinzeller and A. Kotyk (eds.), Acad. Press, New York 1961, pp. 87-93.

KIMMICH, G.A.: Preparation and properties of mucosal epithelial cells isolated from small intestine of the chicken. Biochemistry 19, 3659-3668 (1970a).

KIMMICH, G.A.: Active sugar accumulation by isolated intestinal epithelial cells. A new model for sodium-dependent metabolite transport. Biochemistry 19, 3669-3677 (1970b).

KIPNIS, D.M., PARRISH, J.E.: Role of $Na^+$ and $K^+$ in sugar (2-deoxyglucose) and amino acid ($\alpha$-aminoisobutyric) transport in striated muscle. Fed. Proc. 24, 1051-1059 (1965).

KLEINZELLER, A., KOTYK, A.: Cations and transport of galactose in kidney-cortex slices. Biochim. Biophys. Acta 54, 367-369 (1961).

KROLENKO, S.A., ADAMJAN, S.J.: Permeability of muscle fibers to nonelectrolytes. Cytologia 9, 185-192 (1967).

KROMPHARDT, H., GROBECKER, H., RING, K., HEINZ, E.: Über den Einfluß von Alkali-Ionen auf den Glycintransport in Ehrlich-Ascites-Tumorzellen. Biochim. Biophys. Acta 74, 549-551 (1963).

OXENDER, D.L.: Stereospecifity of amino acid transport for Ehrlich ascites tumor cells. J. Biol. Chem. 240, 2976-2982 (1965).

PIETRZYK, C., HEINZ, E.: Present symposium (1972).

POTASHNER, S.J., JOHNSTONE, R.M.: Cations, transport and exchange diffusion of methionine in Ehrlich ascites cells. Biochim. Biophys. Acta 203, 445-456 (1970).

POTASHNER, S.J., JOHNSTONE, R.M.: Cation gradients, ATP and amino acid accumulation in Ehrlich ascites cells. Biochim. Biophys. Acta 233, 91-103 (1971).

RIGGS, T. R., WALKER, L. M., CHRISTENSEN, H. N.: Potassium migration and amino acid transport. J. Biol. Chem. $\underline{233}$, 1479-1484 (1958).

RIKLIS, E., QUASTEL, J. H.: Effects of cations on sugar absorption by isolated surviving guinea pig intestine. Can. J. Biochem. Physiol. $\underline{36}$, 347-362 (1958).

ROSENBERG, T., WILBRANDT, W.: Uphill transport induced by counterflow. J. Gen. Physiol. $\underline{41}$, 289-296 (1957).

SCHAFER, J. A., HEINZ, E.: The effect of reversal of $Na^+$ and $K^+$ electrochemical potential gradients on the active transport of amino acids in Ehrlich ascites tumor cells. Biochim. Biophys. Acta $\underline{249}$, 15-33 (1971).

SCHAFER, J. A., JACQUEZ, J. A.: Change in $Na^+$ uptake during amino acid transport. Biochim. Biophys. Acta $\underline{135}$, 1081-1083 (1967).

SCHAFER, J. A., JACQUEZ, J. A.: Evidence against the sodium-gradient hypothesis for amino acid transport in the Ehrlich ascites tumor cell. Fed. Proc. $\underline{27}$, 516 (1968).

SCHULTZ, S. G., CURRAN, P. F.: Coupled transport of sodium and organic solutes. Physiol. Rev. $\underline{50}$, 637-718 (1970).

SIEBERT, G., LANGENDORF, H., HANNOVER, R., NITZ-LITZOW, D., PRESSMAN, B. C., MOORE, C.: Untersuchungen zur Rolle des Natrium-Stoffwechsels im Zellkern der Rattenleber. Z. Phys. Chem. $\underline{343}$, 101-115 (1965).

STEIN, W. D.: The movement of molecules across cell membranes. Academic Press, New York 1967, pp. 177-241.

VIDAVER, G. A.: Transport of glycine by pigeon red cells. Biochemistry $\underline{3}$, 662-667 (1964a).

VIDAVER, G. A.: Glycine transport by hemolyzed and restored pigeon red cells. Biochemistry $\underline{3}$, 795-799 (1964b).

VIDAVER, G. A.: Mucate inhibition of glycine entry into pigeon red cells. Biochemistry $\underline{3}$, 799-803 (1964c).

VIDAVER, G. A.: Some tests of the hypothesis that the sodium-ion gradient furnishes the energy for glycine-active transport by pigeon red cells. Biochemistry $\underline{3}$, 803-808 (1964d).

WHEELER, K. P., CHRISTENSEN, H. N.: Role of $Na^+$ in the transport of amino acids in rabbit red cells. J. Biol. Chem. $\underline{242}$, 1450-1457 (1967).

TABLE I

The Effect of the Potassium Electrochemical Potential Gradient on the Influx of AIB

| Exp. No. | $\Delta\tilde{\mu}_N$ | $-\Delta\tilde{\mu}_K$ | $\Sigma X_+$ | $J_{AIB}$ |
|---|---|---|---|---|
| | | joules mole$^{-1}$ | | $\mu$moles/g/3 min |
| 14 | -880 | -801 | -1681 | -0.34 |
| 30 | -756 | +3438 | +2682 | +5.42 |
| 8a | -563 | -769 | -1333 | +4.27 |
| 18 | -550 | -405 | - 965 | -0.76 |
| 24 | -584 | +4023 | +3439 | +8.86 |
| 16 | -688 | -302 | - 990 | -1.55 |
| 19 | -385 | -797 | -1182 | +0.31 |
| 10a | -225 | -385 | - 610 | +3.84 |
| 23 | -117 | +4028 | +3911 | +7.97 |
| 45 | -1267 | -1034 | -2301 | +0.82 |

The above data is taken from experiments reported by Schafer and Heinz (1971) in which the $Na^+$ electrochemical potential gradient ($\Delta\tilde{\mu}_N$) was approximately the same, but in which the potassium electrochemical potential gradient ($\Delta\tilde{\mu}_K$) showed wide variation. $J_{AIB}$ is the three-min influx of AIB per g dry weight of ascites cells, measured at an extracellular AIB concentration of approx. 3.5 mM. The correlation coefficient for $J_{AIB}$ plotted as a function of $\Delta\tilde{\mu}_N$ was 0.41, whereas that for $J_{AIB}$ plotted as a function of $\Sigma X_+$, the sum of the $Na^+$ electrochemical potential gradient and the negative of the $K^+$ electrochemical potential gradient, was 0.85. To compute the ion electrochemical potential gradients, the transmembrane potential was estimated by the chloride distribution. Since the extracellular AIB concentrations were almost the same in all experiments listed, the AIB chemical potential gradient was not included in the summation of the gradients.

# Some Observations on the Nonhomogeneous Distribution inside the Ehrlich Cell

C. Pietrzyk and E. Heinz
Institut für vegetative Physiologie der Universität Frankfurt/Main, Germany

There is a little doubt that the active transport of amino acids in many animal tissues may utilize energy from the electrochemical potential (ECP) gradients of $Na^+$ and $K^+$. It is highly uncertain, however, whether this source of energy is sufficient to meet the total requirement of this transport. In several systems it has been observed that during full metabolic activity higher accumulation ratios are obtained than during metabolic inhibition, even with the same Na and K gradients present (1, 2). As we have heard, Schafer and Heinz recently found that, of the minimum energy necessary for AIB transport in Ehrlich cells, about 4, 000 joules per mole is not accounted for by the $Na^+$ and $K^+$ gradients (3); this deficit represents more than half of the minimum energy necessary. The question arises whether this apparent deficit is supplied directly by metabolism, via a chemi-osmotic coupling, as in primary active transport. This would mean that the AIB transport is driven only in part by the ECP gradients of the electrolyte ions and, in the other part, by direct coupling to a chemical reaction, e. g. to the cleavage of energy-rich phosphate bonds. To devise an appropriate model for such a twofold coupling is difficult, because the two couplings do not appear to act side-by-side, independently of each other. For example, in the absence of $Na^+$ in the medium, amino acids are hardly transported at all, as if even the "primary" transport of AIB depended on Na ions (4). Before invoking a direct chemi-osmotic coupling, we should exclude all other possibilities. One such possibility would be that the gradients Schafer and Heinz used in their calculations are too low. These gradients had been derived from the intra- and extracellular concentrations of Na and K ions, a procedure based on the assumptions that the cations are evenly distributed over the total cellular space and that their activity coefficients are the same inside and outside the cell. If, however, there were cellular compartments, each with a different Na concentration, the true ion gradients across the cellular membrane might be quite different than inferred above. A special kind of compartmentalization has been proposed by Eddy, who, in an attempt to explain the difference in transport activity between respiring and inhibited cells, suggested that, owing to the high activity of the sodium pump, a Na-depleted region develops at the inner face of the cellular membrane (2). As a consequence, the effective $Na^+$ gradient, i. e. that between the extra-cellular $Na^+$ and that in the Na-depleted region, would be much steeper than it would be with an even $Na^+$ distribution throughout the cells. For K ions the reverse might hold. During metabolic inhibition clearly no such Na-depleted region could develop as the Na/K pump is paralyzed. This hypothesis seems plausible, but quantitative considerations have led us to abandon it. The hypothesis is based on two tacit assumptions:

1) The amino acid, contrary to the Na (and K )ions, must distribute evenly over the cellular space. In other words, the amino acid must be maintained at a ligher level in the Na-depleted region than in other parts of the cell. Otherwise additional energy would be required to overcome this concentration barrier, so that the energy gained through the Na-depleted region would be, at least partly, expended to meet this extra requirement.

2) The Na-depleted region, which can only be <u>transient</u>, and should vanish in the steady state, must be maintained long enough to be effective. Considering the incubation times for amino acid uptake in Ehrlich cells, usually longer than 1 min, and considering the time necessary to equilibrate the cell with its surrounding, it would be necessary to suppose that the $Na^+$ concentration in the above region is maintained at a sufficiently low level for at least 5 minutes.

As to the first assumption, it seems a priori unlikely since it implies either that the mobility of the amino acid in the cytoplasm is much higher than that of the alkali ions, or that there exists a specific barrier for these ions which lets amino acids pass freely.

As to the second assumption, a few calculations, based on reasonable premises, will demonstrate that in the absence of true barriers any difference in ion concentrations would have completely disappeared after 5 minutes. To keep the calculations simple, we assume that the pump in the cellular membrane operates at a sufficient rate to maintain an approximately constant $Na^+$ gradient across the cellular membrane. We assume that this pump rate is high enough to produce this gradient before the $Na^+$ inside the cell has come to equilibrium. The concentration profile of the $Na^+$ inside the cell would now behave as in a homogeneous membraneless sphere suspended in a medium with a $Na^+$ concentration as low as that assumed to prevail in the region near the inner face of the cell membrane. A simplified equation to calculate the decline of the intracellular concentration gradients owing to diffusion is given by Jost (5).

$$\frac{\bar{C}_t - C_{oo}}{C_i - C_{oo}} = \frac{6}{\pi^2} \exp\left(-\frac{\pi^2 Dt}{r^2}\right)$$

$C_t$ is the average cellular concentration at time t. $C_i$ is the initial concentration of the sphere. $C_{oo}$ is the final concentration reached after all concentration differences inside the cell have vanished. D is the diffusion constant in $cm^2/sec$; $\underline{t}$ is the time in sec, $\underline{r}$ is the radius of the cell in cm.

The radius of the cells can, with fair approximation, be taken to be $10\mu = 10^{-3}$ cm. The diffusion coefficient for NaCl and KCl in free solution is about $1.5 \cdot 10^{-5}$ $cm^{-2}/sec$. The corresponding diffusion coefficient inside the cell according to more recent work, is about half that value (6). To be on the safe side, we have assumed that inside the cell $D = 0.3 \cdot 10^{-5}$ $cm^2/sec$, i.e. about 1/5 of the value in free solution. Using this simplified formula we can estimate the times at which the left-hand ratio has become 1/2, 1/10, or 1/100 its initial value.

TABLE I

Diffusion within a homogeneous sphere (see text)

| $\dfrac{\bar{C}_t - C_{oo}}{C_o - C_{oo}}$ | per cent equilibration | time m sec |
|---|---|---|
| 0.5 | 50 | 7 |
| 0.1 | 90 | 61 |
| 0.01 | 99 | 139 |
| 0.001 | 99.9 | 217 |

We see that after 7 mseconds the left-hand ratio is only 50%, and that after 0.2 seconds, less than 0.1% its initial value. In other words, after much less than one second the unequal distribution inside the cell has almost completely disappeared. In order for the halftime of equilibration to be of the order of 5 min or higher, we would have to assume the intracellular diffusion constants to be many orders of magnitude smaller than those in free solution, contrary to all evidence. Hence for the cytoplasm we can safely assume that after a few seconds there is not any $Na^+$-depleted region left. If the pump is less vigorous than assumed, the equilibration would have been completed still earlier.

The situation may be different if a true barrier is involved, surrounding a compartment and preventing a rapid equilibration of $Na^+$ over the total cellular space. Such a barrier might be in the nuclear membrane. Since in Ehrlich cells the nuclei occupy one third or more of the cellular space, any unequal distribution between nuclei and cytoplasm might cause considerable error in estimating the true Na-gradient across the cell membrane. It has repeatedly been reported that normally the nuclei have a higher $Na^+$ concentration than the cytosol, and it has been suggested that the $Na^+$ concentration of the nuclei equilibrates with the outside medium faster than with the cytosol (7). Hence at low $Na^+$ in the medium the nuclear $Na^+$ concentration might be lower than the overall cellular $Na^+$ concentration. Thus, either with high or with low cellular $Na^+$ we are bound to derive an erroneous $Na^+$ gradient, although the error should in each case be in a different direction. We have therefore started to isolate nuclei from Ehrlich cells and to determine their $Na^+$ content relative to that in the cytoplasmic fraction of the cell. In order to prevent the nuclei from losing $Na^+$ during the procedure, we exclusively used non-aqueous solvents, following the method originally developed by Behrens and modified by Siebert (8, 9). Freeze-dried cells were suspended in petrolether and ground by glass beads for several hours, their gradual disintegration being followed in microscopical samples. The ground material, containing nuclei and nuclear fragments, was fractionated with respect to specific gravity by centrifugation in mixtures of cyclohexane and carbon tetrachloride.

The DNA was determined according to Schneider (10), and the alkali ions by flame photometry. In addition to pure nuclei we obtained also fractions with nuclear fragments whose amount was estimated by DNA determination, taking advantage of the fact that DNA is almost exclusively localized in the nucleus. Plotting the $Na^+$ content versus the DNA content of each fraction (Fig. 1), the former is seen to rise with the latter, as would be expected if the nuclei contain more $Na^+$ than the cytosol. The values at the far right hand correspond to almost pure nuclei and those at the far left hand, to pure cytoplasm.

The true cytoplasmic $Na^+$ can now be estimated by correcting the total cellular Na for the extra $Na^+$ associated with DNA, which can be obtained from the slope $\Delta$ Na/ $\Delta$ DNA. The results obtained by this procedure in two typical experiments are shown in table II. With normal $Na^+$ in the medium the "true" cytoplasmic $Na^+$ was found to be about half the overall cellular $Na^+$ (Table II, upper line). Assuming that the electrical p.d. either does not contribute to the driving force or is not subject to a major correction, and considering the almost even distribution of K ions between cytoplasm and nucleic material, the correction of the cytoplasmic $Na^+$ corresponds to an extra driving force of about 1800 joules/mole. It should be remembered that the deficit in driving force is at least 4000 joules/mole.

A possible error in the above calculations might result from an admixture to the separated fractions of extracellular Na which is trapped with the centrifugation of

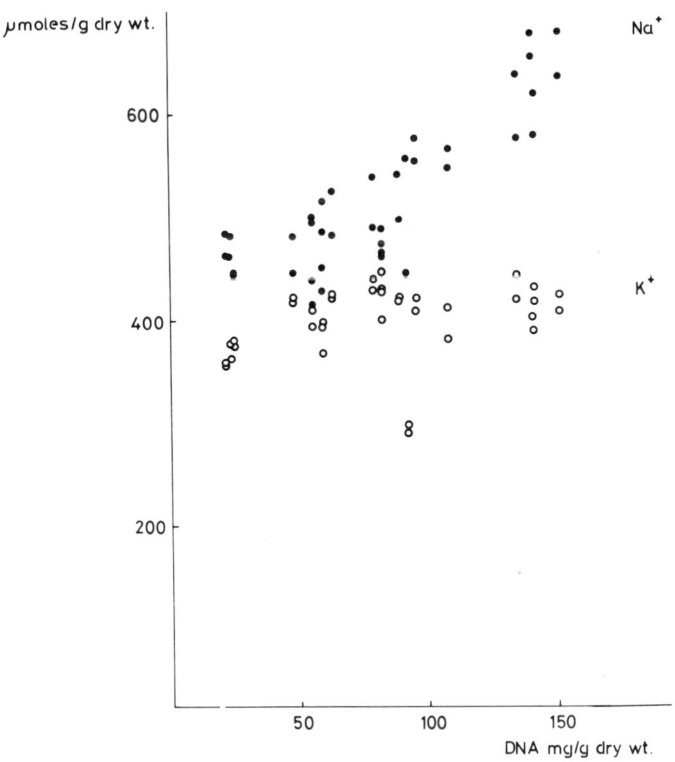


Fig. 1. Distribution of Na$^+$ and K$^+$ in the cell as a function of DNA. The Na$^+$ and K$^+$ content of each cellular fraction is plotted versus the DNA content. Assuming a linearity we obtain functions with the following regression lines, calculated by the method of least squares:

$$\text{for Na}^+ : y = 400 + 1.42\ x \quad \mu\text{moles} \cdot g^{-1}$$
$$\text{for K}^+ : y = 385 + .22\ x \quad \mu\text{moles} \cdot g^{-1}$$

The slope of the Na$^+$ line is significantly different from zero ($P < 0.001$); that of the K$^+$ line is not (n. s.). The slopes are taken to represent the extra ions sequestered in the nuclei per mg DNA

the unbroken cells. That is, if this trapped Na were more abundant in the nonnuclear than in the nuclear fragments of the cells, then the extra Na associated with DNA would appear lower than it actually is. In order to assess this possible error, the cells were once quickly suspended in a cold solution containing labeled inorganic phosphate before the drying and fractionation procedure. It had been previously tested that inorganic PO$_4$ does not enter the cells significantly under these conditions. The labeled phosphate appeared fairly evenly in all fractions, indicating that the contamination by extracellular solutes is rather equal with all fractions. Hence it can be safely assumed that in these experiments the admixture of Na is rather equal with all fractions, and that the above calculations cannot be very much in error.

It has been mentioned that after incubation with low Na no extra Na is associated with the DNA, so that the cytosol Na can be taken as equal to the overall cellular Na concentration. Hence the ECP gradient calculated for the cells incubated in low Na

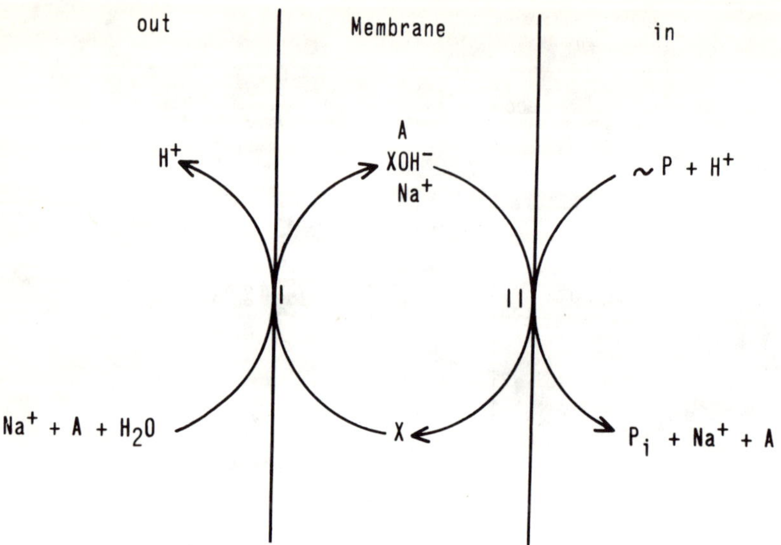



Fig. 2. Coupling model involving a "proton translocator ATPase". The heterophasic hydrolysis of the energy rich phosphate (E~P) proceeds in 3 steps:

1. $Na_o^+ + A_o + H_2O_o + X \longleftrightarrow (X \cdot OH^- \ Na^+A)_o$
2. $(XOH^- Na^+A)_o \longleftrightarrow (XOH^- Na^+A)_i$
3. $(XOH^- Na^+)_i + E{\sim}P_i + H_i^+ \longleftrightarrow E_i + P_i + Na_i^+ + A_i + X$

overall: $Na_o^+ + A_o + H_2O + E{\sim}P + H_i^+ \longleftrightarrow E + P_i + Na_i^+ + A_i + H_o^+$

must be correct. In the same experiments at low extracellular Na, the K was found to be higher in the nuclear fractions than in the nonnuclear ones. Hence the true ECP gradient of K in such experiments must have been lower than the one calculated by Schafer and Heinz. It can be concluded that in Schafer and Heinz' experiments with inversed $Na^+$ gradient the cytoplasmic Na was equal to the overall cellular Na whereas the cytoplasmic K was even lower, and hence less favorable to amino acid transport, than had been anticipated by these authors. In summary, the present investigations appear to show that in experiments with low extracellular Na the effective ECP gradient of this ion across the cellular membrane is correctly obtained from the distribution of Na between cell and medium. Only in the experiments with high extracellular Na may the enrichment of the nucleus with Na tend to make the ECP gradient of Na across the cellular membrane appear smaller than it really is. The resulting error, however, seems far too small to account for the energy deficit estimated by Schafer and Heinz. The behaviour of K does not make much difference or may make the situation even worse. Even though the present experiments are preliminary in that not all sources of error are reliably excluded, they do not seriously detract from the validity of Schafer and Heinz' calculations, and hence do not support the view that the transport of AIB in Ehrlich cells be fully accounted for by the alkali ion gradients. In other cells whose nuclei occupy a much smaller part of the cellular volume than do those of the Ehrlich cells, the unequal distribution of Na between nucleus and cytoplasm would be even less significant.

Another possible source of error in estimating the ECP gradient of Na across the cell membrane may be connected with the intracellular activity coefficient of $Na^+$. As has been pointed out before, the calculations of Schafer and Heinz imply that the activity coefficient of Na is approximately the same in the medium and in the cytoplasm. Dr. Jacquez has emphasized before that the activity of Na inside the Ehrlich cell is not yet known. In view of activity measurements with other cells, however, it appears to be highly unlikely that in Ehrlich cells the cytoplasmic activity coefficients of K and Na are very different from the extracellular ones. I may quote recent results by Pfister et al. who potentiometrically determined the mean activity coefficient of KCl in a reconstituted cytoplasmic medium obtained from liver cells and erythrocytes and found it to be at least 90% that of the external medium (11). These measurements should certainly be carried out with Ehrlich cells, but the prospect of getting much different results appears to be very dim. So far we have little reason to assume that the source of error due to overestimating the cytoplasmic activity coefficient be significant.

As long as the direct coupling between amino-acid transport and a metabolic reaction is not established, it seems premature to devise models of it. Considering, however, that the amino-acid transport may be coupled simultaneously to a metabolic reaction and to an electrolyte ion gradient, it appears justified to ask how such double coupling could work. Any model to visualize it ought to take into account that Na ions are required for both kinds of coupling. I have tried to adjust P. Mitchell's model of "ATPase I" (12) to meet this requirement (Fig. 2). This modified model is supposed to tie both couplings into a single overall reaction, which requires both Na ions and an energy-rich phosphate for maximal activity. Even though it may account for many observations, it decisively depends, as does its original, on the feasibility of a "heterophasic" reaction in which $\sim P$ on the right side is hydrolysed using $OH^-$ from the left side of the membrane. For the phosphotransferase system of sugar transport a heterophasic reaction, the phosphorylation of the sugar, has been shown to occur (13). Whether a hydrolytic reaction can also proceed heterophasically, as is postulated in the present model, is highly uncertain.

# References


1. VIDAVER, G. A.: Biochemistry 3, 799 (1964).
2. EDDY, A. A.: Biochem. J. 108, 489 (1968).
3. SCHAFER, J. A., Heinz, E.: Biochim. Biophys. Acta 249, 15 (1971).
4. HEINZ, E.: in "Permeability and Function of Biological Membranes", L. Bolis (ed.)
5. JOST, W.: Diffusion, Academic Press, New York 1960, p. 45
6. KUSHMERICK, M. T., PODOLSKY, R. T.: Science 166, 1297 (1969).
7. SIEBERT, G., LANGENDORF, H., HANNOVER, R.: Hoppe-Seylers Z. Physiol. Chem. 343, 101 (1965).
8. BEHRENS, M.: in "Biochem. Taschenbuch", H. Rauen (ed.), Springer, Heidelberg 1956, p. 910.
9. SIEBERT, G.: Biochem. Z. 334, 369 (1961).
10. SCHNEIDER, W. C.: J. Biol Chem. 164, 747 (1946).
11. PFISTER, H.: Z. Naturforsch. 25 b, 1130 (1969).
12. MITCHELL, P.: in "The Molecular Basis of Membrane Functions", D. C. Tosteson (ed.), Prentice Hall, Inc., Englewood Cliffs 1969, p. 483.
13. KABACK, H. R.: J. Biol. Chem. 243, 3711 (1968).

## TABLE II

Cytoplasmic $[Na^+]$ and $[K^+]$ are obtained from overall cellular contents of these ions after subtraction of the extra equivalents of ions presumably sequestered in the nuclei. These equivalents are obtained from the slopes of the corresponding regression lines, as illustrated in Fig. 1, presuming that the contamination with extracellular ions is uniform in all fractions (see text). Slopes that are small and insignificant (n. s.) are taken to be zero. The extracellular $K^+$ in both experiments was about 8 mM.

| | Medium | | | Cells | | | | $\Delta \Sigma \nu X$ |
|------|--------|----------------------------|---------------------------|---------|-----------|---------|-----------|------------------------|
| Exp. | $[Na^+]$ | $\dfrac{\Delta\,Na^+}{\Delta\,DNA}$ | $\dfrac{\Delta\,K^+}{\Delta\,DNA}$ | $[Na^+]$ overall | cytoplasm | $[K^+]$ overall | cytoplasm | Gain in vectorial driving force |
| No. | mM | $\mu$ eq/mg | $\mu$ eq/mg | mM | mM | mM | mM | joules/mole |
| 11 | 146.5 | 1.42 | n. s. | 64.2 | 36.8 | 107.0 | 107.0 | +1420 |
| 12 | 25.3 | n. s. | 0.91 | 30.9 | 30.9 | 149.8 | 126.7 | - 410 |

# The Influence of H⁺, Na⁺ and K⁺ on the Influx of Glutamate in Ehrlich Ascites-Tumor Cells


P. Geck and B. Pfeiffer
Institut für vegetative Physiologie der Universität Frankfurt/Main, Germany



## Summary

The concentration dependence of glutamate uptake is described as the sum of a saturable (mechanism I) and a nonsaturable (mechanism II) component. Lowering the pH does not influence mechanism II, but stimulates mechanism I.

$Na^+$ and $K^+$ activate both mechanisms at neutral pH. No influence of $Na^+$ and $K^+$ on the uptake was observed at low pH.

The transport of glutamate is inhibited by neutral amino acids. As the pH increases the inhibition of mechanism II increases, while for mechanism I the inhibition decreases.


## Introduction

Earlier investigations have shown that Ehrlich ascites tumor cells transport glutamate actively. There is no saturation of the transport by concentrations up to 20 mM. At very high concentrations of glutamate ($K_m \approx 200$ mM), however, it is saturable. Many neutral amino acids inhibit strongly. Inhibition by acidic amino acids was not observed. In contrast to the uptake of neutral amino acids, which is inhibited by an increasing proton concentration, the influx of glutamate is stimulated as the pH decreases.

Based on the above-mentioned findings (1, 2, 3), this paper reports investigations, which characterise further the glutamate transport in Ehrlich ascites tumor cells.

## Methods

The preparation of the cell suspension and incubation procedure were described before (4).

All incubations were made at 37°C.

Krebs-Ringer phosphate buffer containing 1% albumin was used as medium. When the concentrations of the alkali ions were varied, isoosmolarity was maintained by choline.

## Results

In Ehrlich ascites tumor cells the dependence of glutamate uptake on the concentration of glutamate in the investigated region from 0.01 up to 20 mM can be described as the sum of a saturable and a nonsaturable portion, so that following equation describes the influx:

$$J = J_{Max} \cdot \frac{[Glu]}{[Glu] + K_m} + k \cdot [Glu]$$

The saturable fraction, for which the Michaelis–Menten relation is valid, shall be called mechanism I in the following, the fraction nonsaturable up to 20 mM be mechanism II.

The relative influx (J/[Glu] ) at two pH levels is shown in Fig. 1. Obvious is the

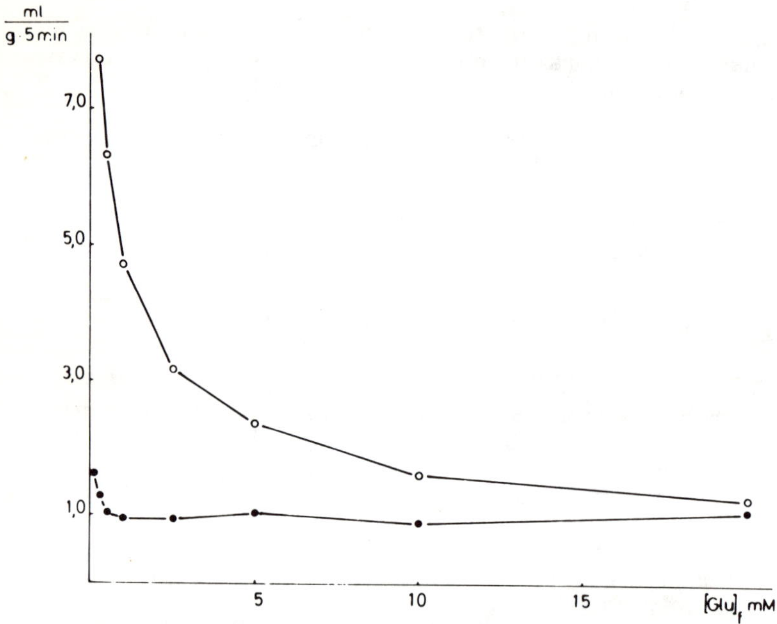


Fig. 1. Relative glutamate influx at two pH values. o pH 5.4; ● pH 7.25

activation of mechanism I by protons. In the investigated region from pH 7.2 to 5.2 lowering the pH causes $J_{max}$ and $K_M$ to increase. $J_{max}$ increases more than $K_M$, so that $J_{max}/K_M$ also increases as the proton concentration increases. Protons do not influence mechanism II. The value of $J_{max}$, $K_M$, $J_{max}/K_M$ and k as determined by various experiments are shown in Fig. 2. The pK of the group responsible for the proton effect is at a pH below 5 as the figure shows. The influence of $Na^+$ and $K^+$ on the uptake of glutamate was investigated at a pH of about 5 and of about 7.2. At the lower pH the results show no influence of either $Na^+$ or $K^+$ on the two mechanisms (Fig. 3). At neutral pH, mechanism II is activated by $Na^+$ and $K^+$ (Fig. 4, Fig. 5) but $K^+$ concentrations above 30 mM act inhibitory. The two ions seem to have no noticeable influence on each other's bindings. Both are required to produce a significant flux. In neutral medium, mechanism I seems to be activated by $Na^+$ and $K^+$ as well, since the exclusive activation of mechanism II cannot explain the extent of activation at 0.01 mM glutamate. At pH 7.2 the following constants for activation and inhibition by $Na^+$ and $K^+$ are found for mechanism II:

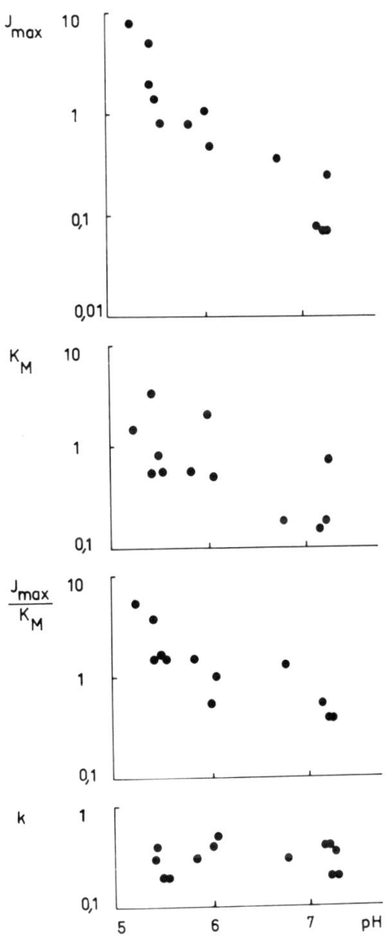



Fig. 2. pH dependence of $J_{max}$, $K_M$, $J_{max}/K_M$ and k

| | |
|---|---|
| activation by Na$^+$ | K $\approx$ 4 mM |
| activation by K$^+$ | K $\approx$ 0.25 mM |
| inhibition by K$^+$ | K $\approx$ 200 mM |

In neutral as well as acidic medium, the glutamate uptake is inhibited by a number of neutral amino acids. The mode of inhibition however is different in the two regions. The inhibition by glycine and glutamine was investigated more extensively. At pH 7.2 mechanism I is not influenced by the two amino acids, while mechanism II is strongly inhibited (Fig. 6). The $K_i$ of glycine is about 3 mM and in good agreement with the $K_M$ of glycine influx (5). In acidic medium on the other hand mechanism I is competitively inhibited by both amino acids, while mechanism II remains uninfluenced (Fig. 7). The pH dependence of the inhibition by glycine and glutamine was investigated at a low (0.01 mM, mechanism I) and a high (20 mM, mechanism II) concentration of glutamate. Both amino acids show a similar behaviour. In the case of mechanism I the inhibition decreases, while for mechanism II the inhibition increases as the pH increases (Fig. 8). The value of pK for the pH dependence of inhibition is about equal for both mechanisms and for both amino acids. It is found between 6 and 7.

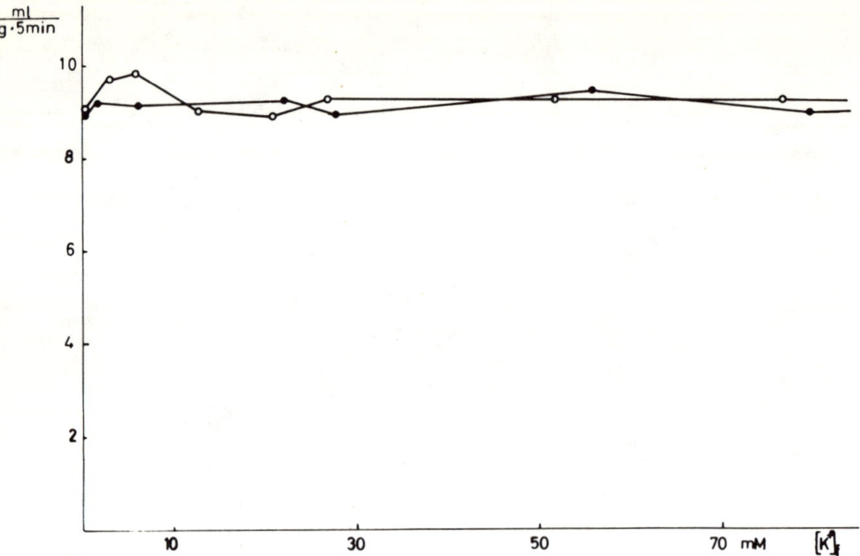

Fig. 3. Effect of K$^+$ on the relative glutamate influx at two Na$^+$ concentrations.
pH 4. 9     [Glu] = 2 mM     ● 4 mM Na$^+$     o 44 mM Na$^+$

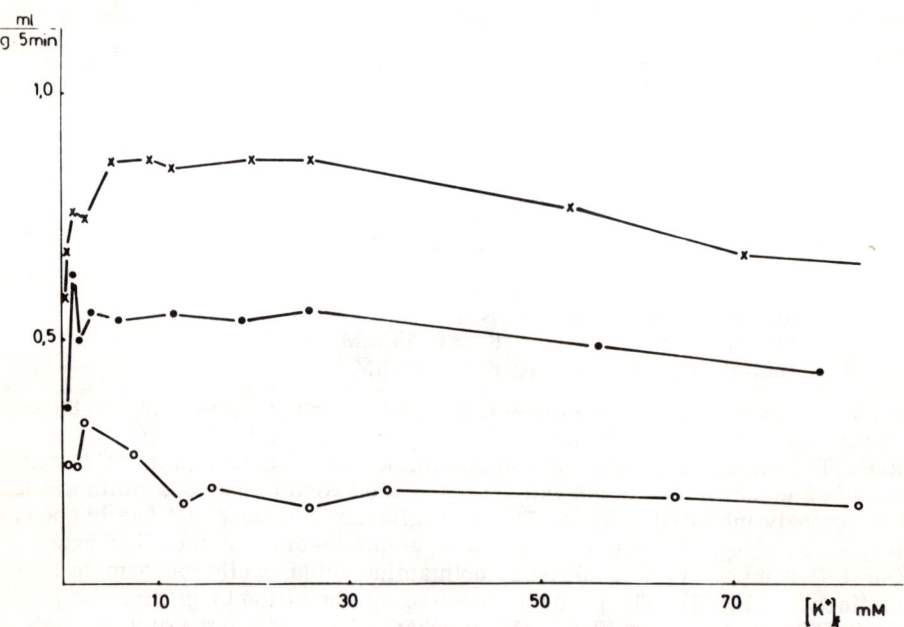

Fig. 4. Effect of K$^+$ on the relative glutamate influx at three Na$^+$ concentrations
pH 7. 5
x   3. 5 mM Na$^+$     ● 35 mM Na$^+$     o 50 mM Na$^+$

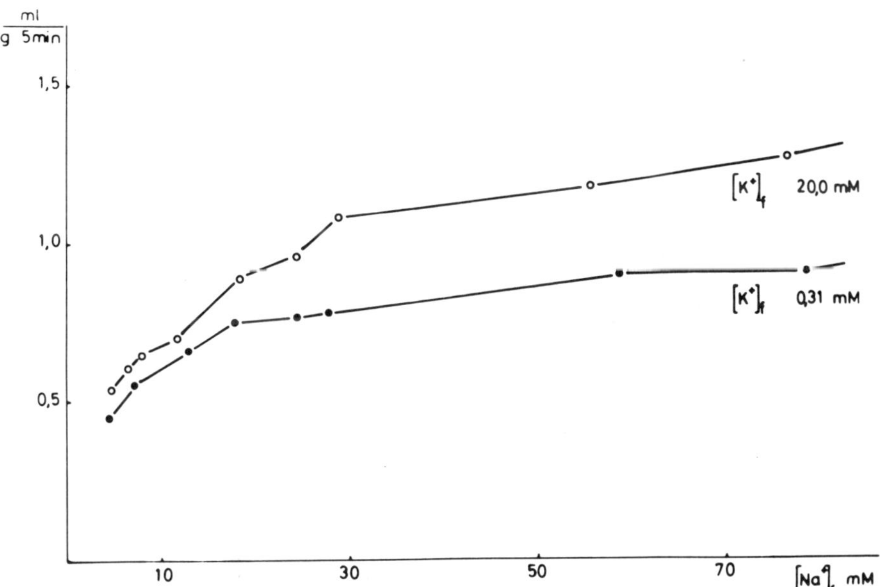



Fig. 5. Effect of Na$^+$ on the relative glutamate influx at two K$^+$ concentrations. pH 7.45

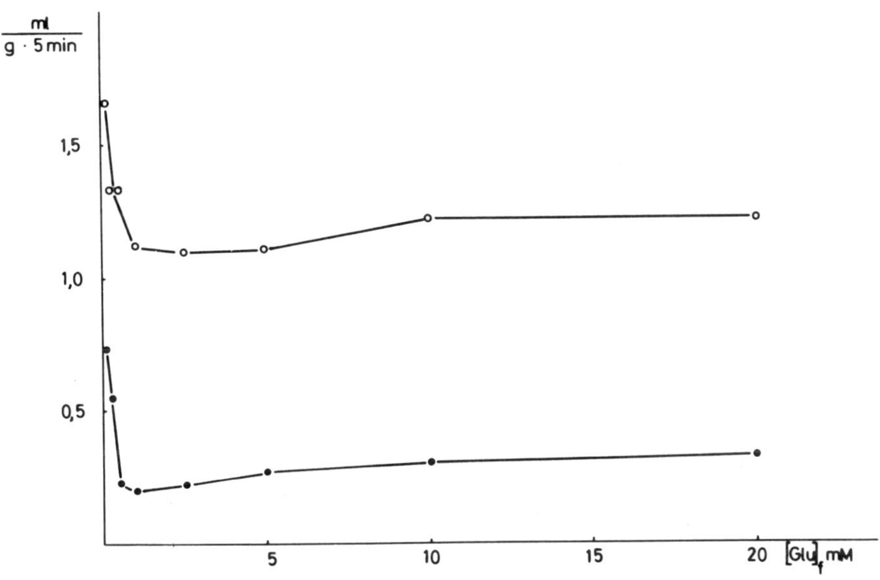



Fig. 6. Inhibition of relative glutamate influx by glycine. pH 7.25
    o    0 mM Gly        ●    20 mM Gly

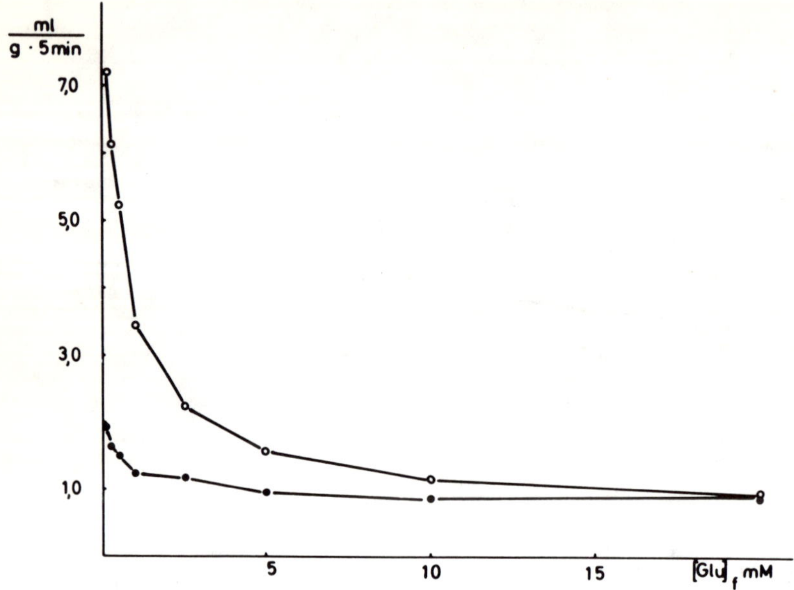

Fig. 7. Inhibition of relative glutamate influx by glycine. pH 5. 55
o    0 mM Gly          ●    20 mM Gly

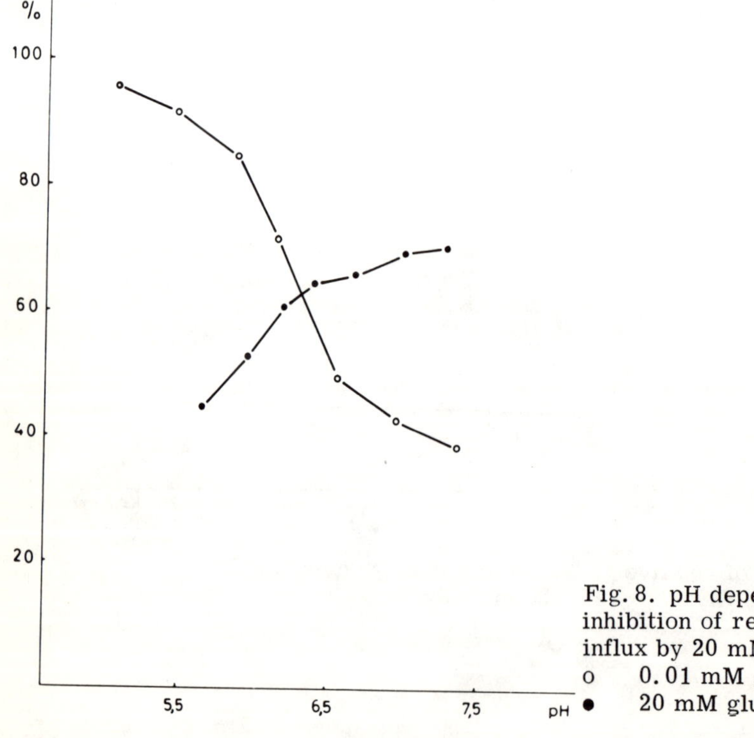

Fig. 8. pH dependence of the
inhibition of relative glutamate
influx by 20 mM glutamine.
o    0. 01 mM glutamate
●    20 mM glutamate

## Discussion

The results show, that the uptake of glutamate by Ehrlich ascites tumor cells is stimulated by $H^+$, $Na^+$ and $K^+$. It is not possible to differentiate whether the ions are bound to the carrier as effectors or act as counter-ions and are transported by another transport mechanism. Since the influence of the ions on the two mechanisms is different, a separate discussion follows.

For mechanism I, protons are the prerequisite of a noticeable flux. They seem to increase the mobility of the carrier substrate complex, but to decrease the affinity of the carrier to the substrate. Since the pK of activation by protons is below pH 5, a carboxyl group may be thought responsible for this effect. The binding of protons to the substrate, which these findings might suggest, does not seem to occur, since then a decrease of $K_M$, not an increase, has to be expected with increasing $H^+$ concentration. This possibility has already been excluded by Heinz et al. (2). In neutral medium $Na^+$ and $K^+$ seem to stimulate the uptake by mechanism I, acting on different sites. Both may be replaced by protons, since they do not stimulate in acidic medium.

Mechanism II does not show an influence of $H^+$ ions on the flux at normal electrolyte concentrations (8 mM $K^+$, 145 mM $Na^+$). Like in mechanism I at neutral pH, $Na^+$ and $K^+$ are necessary for the uptake and do not affect the same site, since both ions are necessary for the transport. They can be replaced by protons at low pH. Contrary to mechanism I, the binding of protons to these sites does not increase the flux.

To characterize the two mechanisms further, the influence of neutral amino acids on both mechanisms was investigated . Inhibition was found in acidic as well as in neutral medium. In the latter mechanism II was inhibited. The inhibition constant of glycine was determined to be about 3 mM, which agrees well with the $K_M$ of glycine transport (5). The inhibition decreases as the proton concentration increases. The pK value of the group responsible lies between 6 and 7. This corresponds to the pH dependence of glycine transport (6). These findings allow us to assume that mechanism II is a transport system of neutral amino acids. Mechanism I is inhibited in acidic medium by neutral amino acids, while in neutral medium no inhibition is found. The pK value of the group responsible for the activation by protons is between 6 and 7. Therefore this group is probably not identical with the one responsible for the activation of glutamate uptake. Since in acidic medium no noticeable flux of glycine is measured, one may suppose a carrier substrate complex with little or no mobility.

The data up to now are still insufficient to suggest a reaction scheme which incorporates all findings. Further investigations of this problem are necessary.

## Acknowledgements

This work was supported by the Deutsche Forschungsgemeinschaft (H 102/10).

## References

1. HEINZ, E., LOEWE, U., DESPOPOULOS, A., PFEIFFER, B.: Biochem. Z. 340, 487 (1964).

2. HEINZ, E., PICHLER, A.G., PFEIFFER, B.: Biochem. Z. **342**, 542 (1965).
3. PICHLER, A.G.: Biochim. Biophys. Acta 104, 104 (1965).
4. GROBECKER, H., KROMPHARDT, H., MARIANI, H., HEINZ, E.: Biochem. Z. 337, 462 (1963).
5. HEINZ, E.: J. Biol. Chem. 211, 781 (1954).
6. KROMPHARDT, H.: Biochem. Z. 343, 283 (1965).

# Evidence for a Sodium-Independent Transport System for Glucose Derived from Disaccharides

W. F. Caspary
Division of Gastroenterology and Metabolism, Department of Medicine,
University of Göttingen, Germany

During the last ten years the number of digestive enzymes bound to the brush border membrane of intestinal mucosal epithelial cells (5, 7, 9) and the number of substrates translocated across the microvillus membrane by $Na^+$-dependent transport systems has increased considerably (2, 5, 9). Evidence is accumulating that digestive and absorptive functions of the small intestine cannot be expressed by the simple concept that enzymatic digestive action and membrane transport are sequential, but otherwise separate processes (5, 7, 8, 14, 15). On the contrary, both processes are closely integrated and are achieving an efficient cooperation, enhanced in addition by spatial organization of the functional components of these processes (5, 6, 7).

To emphasize the significance of the close spatial and functional relationships between brush border hydrolytic enzymatic activities and the intestinal transport system for actively transported sugars Crane introduced the concept of a digestive-absorptive surface (6). On the basis of earlier experiments of Miller and Crane (13) indicating that glucose liberated from sucrose was better transported in the presence of high concentrations of glucose oxidase than glucose liberated from glucose-1-phosphate by the action of membrane-associated alkaline phosphatase, Crane developed a concept of "kinetic advantage" (6) based on the inference that sucrase is more closely organized with transport systems than alkaline phosphatase. This concept is depicted in figure 1. A very similar phenomenon was noted by Parsons in the in vitro frog gut and was given a similar interpretation (14, 15).

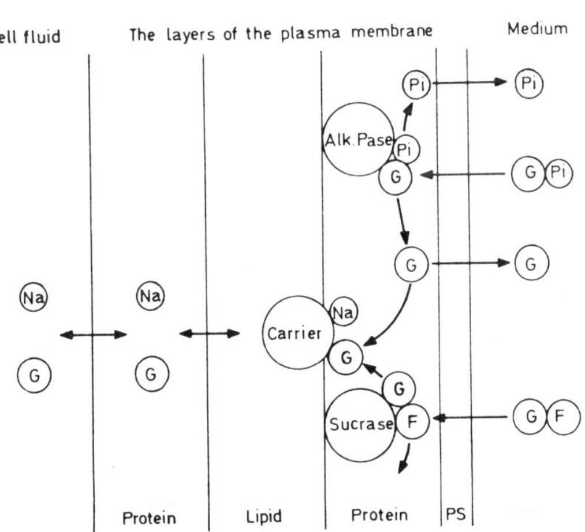


Fig. 1. Presumed structural and functional organization and spatial relationship between brush border hydrolases and transport carriers for glucose in the outer protein coat of the microvillus membrane (from Crane, 1967, ref. 6).
G = glucose, F = fructose, Pi = phosphate, PS = mucopolysaccharide

An alternative proposal has been made to explain the high efficiency of glucose-uptake from sucrose (11), based on the amount and nature of the fuzzy coat, serving as a diffusion barrier for disaccharides into the intermicrovillus spaces, to the disaccharidases, and for liberated glucose into the lumen.

The concept of the "kinetic advantage" helps to explain why glucose moieties from disaccharides are so efficiently absorbed that one can hardly find any glucose released from sucrose in the succus entericus as in human intestinal perfusion studies with sucrose (10) or in the mucosal compartment in in vitro studies (13, 14).

Recently this work was taken up again in an effort to expose more clearly the organizational relationship between sucrose hydrolysis and the transport carriers contributing to the "kinetic advantage". Our experiments to date seem to point to the existence of a new transport system closely associated with membrane-bound disaccharidases and different from the well-known $Na^+$-dependent sugar transport system (4).

Reinvestigation on the nature of the concept of the "kinetic advantage" was taken up again after we had obtained the experimental results depicted in table 1 (Malathi, Caspary, Crane, unpublished). Tissue uptake of 6-deoxy-D-glucose, a non-metabolizable substrate of the active sugar transport system, was measured in the presence of several disaccharides and amylose using segments of hamster small intestine in vitro and 59 mM Tris in the incubation medium to decrease hydrolysis. Tissue uptake of glucose derived from disaccharides and amylose was nearly equal, no matter which substrate had been used, but an inhibitory effect of disaccharides on transport of 6-deoxy-D-glucose was only observed in the presence of maltose and amylose, whereas sucrose and trehalose did not have an inhibitory effect on uptake of 6-deoxy-D-glucose. Analysis of the incubation medium revealed that glucose did appear in the medium to an appreciable extent only with maltose and amylose as substrates in the incubating medium. The most likely explanation is that hydrolysis of maltose and amylose was still exceeding the transport capacity for glucose. These results suggested that at a low rate of hydrolysis uptake of glucose from disaccharides and uptake of 6-deoxy-D-glucose, substrates generally sharing a common $Na^+$-dependent transport pathway (4) might occur in parallel, possibly not interfering with each other. In the case of exceeding hydrolysis, however, glucose may occur in the medium after maximal transport rates are achieved and then might compete with other actively transported hexoses using, like D-glucose, the common sugar transport system.

Using short term incubations (1 or 2 minutes) of segments of hamster small intestine we measured the uptake of glucose from sucrose and free glucose from the medium under conditions close to maximal transport rates for glucose. A typical experiment is given in table 2 (Malathi, Caspary, Crane, unpublished). Using 25 mM $^{14}$C-D-glucose as the substrate and having determined the apparent transport Km for glucose (1.5 mM), we can calculate a maximal transport rate (Vmax) for glucose which in the case of the existence of a single common carrier-mediated sugar transport system should not be exceeded. If glucose from disaccharides is using the same transport system as free glucose does under conditions of nearly maximal transport rates, no additional uptake of glucose should be obtained if additional substrate is introduced from the mucosal compartment. The results obtained, however, show that, if sucrose (15 mM) is added to 25 mM free glucose, the maximal transport rate of glucose can be exceeded by 89% (table 2).

TABLE I

EFFECT OF DIFFERENT DISACCHARIDES ON UPTAKE OF 6-DEOXY-D-GLUCOSE BY HAMSTER SMALL INTESTINE IN VITRO

| Conditions | T/M-ratio ($^3$H-6-DG) | $\mu$moles glucose/ ml tissue water | $\mu$moles glucose/ ml medium |
|---|---|---|---|
| Control (0. 6 mM 6-DG) | 29. 1 | 1. 53 | 0. 08 |
| plus 10 mM Maltose | 6. 75 | 13. 0 | 2. 2 |
| plus 10 mM Sucrose | 28. 1 | 9. 25 | 0. 16 |
| plus 10 mM Trehalose | 29. 9 | 11. 73 | 0. 133 |
| plus 1% Amylose | 18. 8 | 12. 25 | 0. 37 |

Rings of everted hamster small intestine were incubated for 30 min. at 37°C with a modified Krebs-Henseleit phosphate buffer containing 93 mM Na$^+$ and 59 mM Tris. Volume: 5 ml. Substrate: 0. 6 mM $^3$H-6-deoxy-D-glucose

Table II

TISSUE UPTAKE OF FREE GLUCOSE AND GLUCOSE DERIVED FROM SUCROSE
BY HAMSTER SMALL INTESTINE IN VITRO

Segments of everted hamster small intestine were incubated for 1 min. in Krebs-
Henseleit phosphate buffer with substrates as indicated. All data are corrected
for extracellular space. TGO = Tris-glucose-oxidase determination of D-glucose.

| Conditions | $\mu$ moles D-glucose/ml tissue water | |
| --- | --- | --- |
| | TGO | $^{14}$C |
| 15 mM Sucrose | 7.65 | --- |
| 25 mM $^{14}$C-D-glucose | 7.70 | 7.95 |
| 15 mM Sucrose plus 25 mM $^{14}$C-D-glucose | 15.5 | 7.90 |

Calculation: $V_{max} = v \cdot (1 + \frac{K_m}{(S)}) = 7.7 \cdot (1 + \frac{1.5}{25}) = 8.2$

$$8.2 \ \mu \text{moles/ml tissue water} = 100 \ \% \ V_{max}$$

observed: $15.5 \ \mu$ moles/ml tissue water = 189 % $V_{max}$

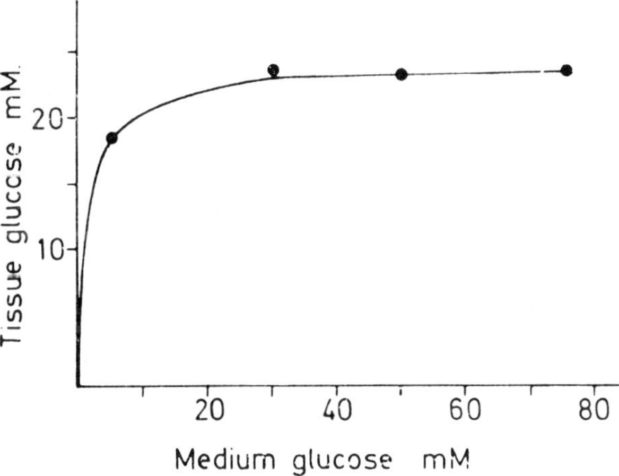

Fig. 2. Concentration-dependent tissue glucose uptake by segments of hamster small intestine. Incubation was for 2 min. in Krebs-Henseleit phosphate buffer at 37°C under pure oxygen

To confirm that nearly maximal loading of tissue with glucose was achieved at concentrations of 20-30 mM of D-glucose,we measured the concentration-dependent uptake of glucose (Fig. 2). Increasing the medium glucose concentration beyond 20-30 mM does not measurably increase tissue uptake of glucose. More systematic studies were done with tissue-loading of glucose from combinations of free glucose and disaccharides as well as glucose-1-phosphate (Fig. 3). When [3]H-D-glucose

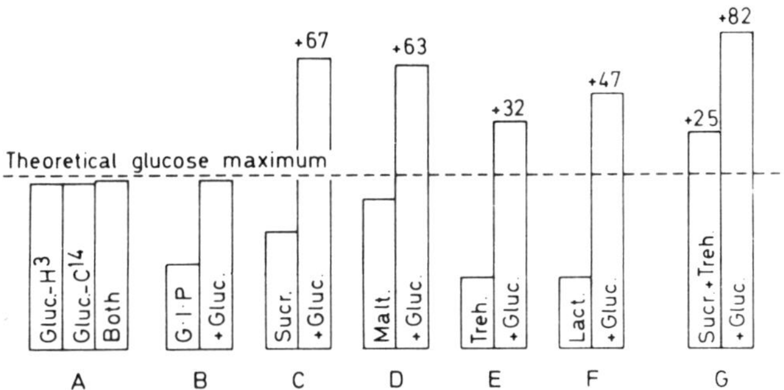

Fig. 3. Tissue loading with glucose from disaccharides and free glucose. Segments of hamster small intestine were incubated for 2 min. in Krebs-Henseleit phosphate buffer with 30 mM [14]C-D-glucose present in all experiments. Additionally are added: 30 mM [3]H-D-glucose (A), 30 mM glucose-1-phosphate (B), 30 mM sucrose (C), 30 mM maltose (D), 30 mM trehalose (E), 30 mM lactose (F) and 30 mM sucrose and 30 mM trehalose (G). The theoretical maximal glucose uptake was calculated from the experimental data of Fig. 2

(30 mM) and [14]C-D-glucose (30 mM) are used independently and together (a total of 60 mM D-glucose),the results obtained (a) are exactly those expected from Fig. 2. However, when any of the disaccharides, sucrose (C), maltose (D), trehalose (E), or lactose (F), is added to 30 mM D-glucose, the component of glucose loading it contributes by itself is substantially additive to the component of glucose contributed by free glucose. Also, when two disaccharides are used in combination (G), their individual components are additive to free glucose, but the glucose component derived from glucose-1-phosphate is not additive to the component of free glucose

(B), thus confirming the earlier observations of Miller and Crane (13).

These results show that mere formation of glucose close to the membrane will not produce the additive effect and permits the assumption of the presence of two kinds of transport systems for glucose.

In the meantime we have additional criteria for the existence of these two kinds of transport systems. Glucose transport in hamster small intestine in vitro is known to depend on the presence of $Na^+$ (4). Glucose loading from sucrose is not or at least not entirely $Na^+$-dependent. Although tissue loading with glucose from sucrose is inhibited in the absence of $Na^+$, almost 50% occurs (Fig. 4). This is not the case

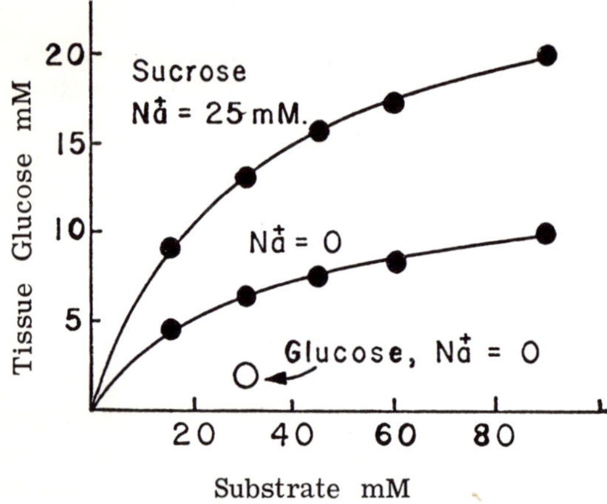


Fig. 4. Glucose uptake from sucrose in absence of $Na^+$. Segments of hamster small intestine were incubated for 2 min. in a modified Krebs-Henseleit phosphate buffer, in which $Na^+$ was replaced to the extent indicated by D-mannitol or choline chloride. (from ref. 8)

for free glucose. Since $Na^+$ has an influence on sucrase activity (18), we do not know at this time whether the low rate of tissue accumulation is due to a decrease of sucrase hydrolytic activity or to inhibition of $Na^+$-dependent absorption of glucose released in excess of the capacity of the sucrase-associated glucose transport system. In any case there is substantial transfer of glucose from sucrose in the absence of $Na^+$. Ramaswamy, Malathi and Crane (16) have shown that a substantial portion of tissue glucose accumulation from additional disaccharides (maltose, isomaltose, trehalose) was independent of added $Na^+$. Under $Na^+$-free conditions individual contributions to tissue glucose were additive when four disaccharides (sucrose, maltose, isomaltose and trehalose) were present together (16). Phlorizin, a competitive inhibitor for the common sugar binding site (1, 4) acted on transport of glucose from disaccharides (sucrose) in a two-fold way (Fig. 5): at low concentrations phlorizin is competitive with a Ki of about $5x10^{-6}M$ as it is with free sugars (1) entering by the $Na^+$-dependent transport system. At high concentrations, however, phlorizin is non- or uncompetitive with a Ki of $10^{-4}M$. This complex inhibition of phlorizin might also be competitive with the existence of two transport systems for glucose from sucrose.

One of the consequences of the $Na^+$-gradient hypothesis and the bifunctional carrier for intestinal glucose transport (4) is, that the affinity of the transport carrier for glucose is increased by $Na^+$ and vice versa the carrier's affinity for $Na^+$ is increased by hexoses of certain minimal structural requirements (3, 4). Actively transported sugars, according to their affinity for the transport carrier, are able to increase the transmural potential difference (PD) across the intestinal tissue and to increase short circuit current (12, 17). If the glucose moiety from disaccharides were to be transported by a different, but $Na^+$-dependent transport system,

Fig. 5. The inhibitory
effect of phlorizin on
tissue glucose uptake
from sucrose. Incuba-
tions were carried out
for 2 min. with substrate
and inhibitor concentra-
tions as indicated. Data
are plotted according to
the method of Dixon.
(from ref. 8)

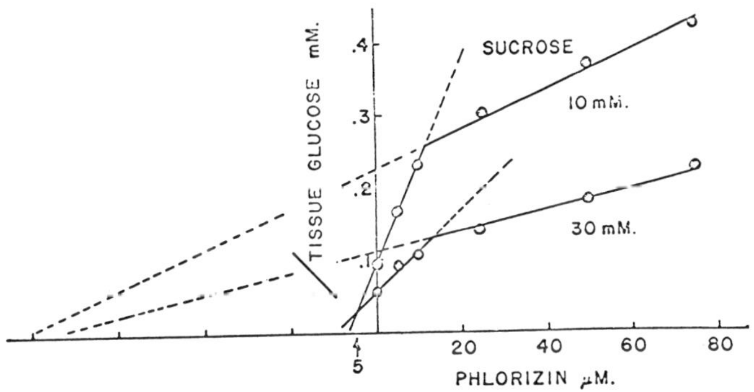


sucrose should induce a further increase in transmural potential difference (PD)
on top of the maximal PD increase induced by free glucose.

As can be seen, however, in Fig. 6, the PD increase evoked by the combination
of glucose (15 mM) and sucrose (15 mM) was not additive, thus indicating that the
amount of additional tissue glucose accumulation was not $Na^+$-sensitive under conditions where the common $Na^+$-dependent glucose transport system was nearly

Fig. 6. Effect of D-gluc-
ose and sucrose on trans-
mural potential difference
(PD) in hamster small
intestine. PD measure-
ments were carried out
as described in ref. 3.
1000 U of glucose oxidase
were added in the experi-
ment indicated to the
mucosal compartment
(15 ml)

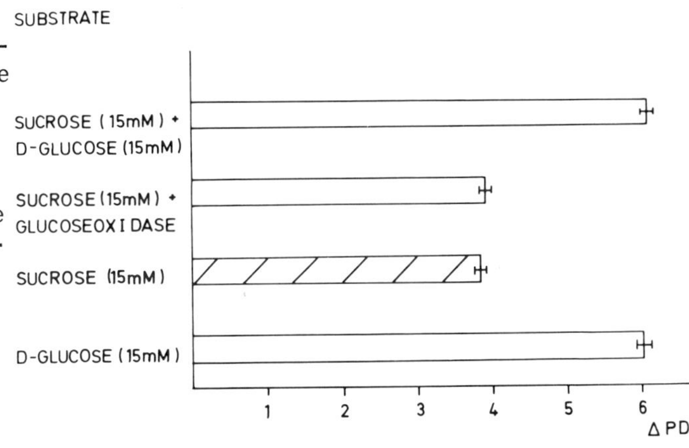


saturated with glucose. In support of the earlier experiments of Miller and Crane
(12) we found that the PD increase induced by glucose from sucrose was not diminished in the presence of glucose oxidase, indicating that glucose oxidase did not
have access to glucose derived from sucrose hydrolysis.

If the potential increase is related to $Na^+$ transfer on a hexose carrier, then glucose from sucrose, if moving on a different carrier, might have been expected to
give an additive effect on PD. The experimental findings that the PD increase
evoked by glucose from sucrose and free glucose was not additive could be consistent either with their using the same carrier or that glucose from sucrose is
using a transport system not generating an increase in transmural PD under conditions of saturation of the PD-generating transport system for free glucose. In
view of the relative $Na^+$-independence of tissue glucose uptake from sucrose and
the measured additive effects of uptake of free glucose and glucose from sucrose
exceeding maximal transport capacities for glucose considerably, the latter interpretation has to be favoured.

That sucrose itself, however, if presented to the mucosal compartment, induced a PD-increase, is explained by the model presented in Fig. 7. As sucrose hydrolysis is exceeding the transport capacity of system B, glucose from sucrose may get access to the potential-generating $Na^+$-sensitive transport system for actively transported hexoses.

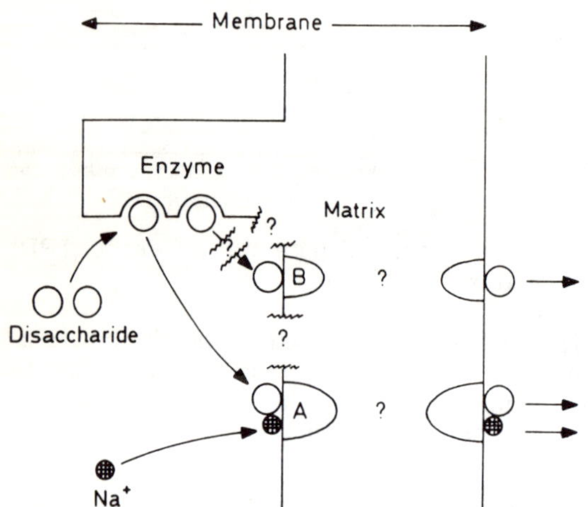


Fig. 7. New concepts of the presumed organization of the microvillus membrane and the structural and functional relationship between brush border disaccharidases and glucose carriers. (from Crane, 1970, ref. 5)

As much as can be assumed from the present data is summarized in Fig. 7. Disaccharidases are integral parts of the outer microvillus membrane. Two different transport systems which might interact at some location of the transport pathway, but having different binding sites, may translocate glucose derived from disaccharide hydrolysis. "A" is the well-known $Na^+$-dependent transport system for actively transported glucologues, evoking a PD increase in the presence of appropriate substrates; "B" is the system preferentially transporting glucose moieties $Na^+$-independently and having a very close relation to the site of hydrolysis. We still do not know the detailed relationship between membrane-associated disaccharidases and system "B", we only know it is a close relationship. We also do not know whether both carriers might interact at some level of the transport pathway. Thus, in view of the experimental evidence, the concept of the "kinetic advantage" of glucose transport from sucrose can be explained by the existence of an additive transport pathway for glucose from disaccharides beeing spatially and functionally closely related to the locus of disaccharide hydrolysis.

Although experimental means are available to differentiate between disaccharide hydrolysis and transport of the glucose moieties from disaccharides (13, 14), the fact that all substrates of brush border disaccharidases exhibited the above phenomenon suggests at least the possibility that these enzymes may of themselves contribute transport function.


## Acknowledgements

The author wants to thank Dr. Robert K. Crane for his guidance and helpful discussions while working in his laboratory on the problem presented in this article. Most of the data presented were obtained in Dr. Crane' s laboratory at Rutgers Medical School, New Brunswick, New Jersey, USA, in close collaboration with my collegue, **Dr. P.** Malathi.

Support was given by the Deutsche Forschungsgemeinschaft (Ca 71/1).


## References

1. ALVARADO, F., CRANE, R. K.: Phlorizin as a competitive inhibitor of the active transport of sugars by hamster small intestine, in vitro. Biochim. Biophys. Acta 56, 170-178 (1962).
2. CASPARY, W. F., CRANE, R. K.: Active transport of myo-inositol and its relation to the sugar transport system in hamster small intestine. Biochim. Biophys. Acta 203, 308-316 (1970).
3. CASPARY, W. F., STEVENSON, N. R., CRANE, R. K.: Evidence for an intermediate step in carrier-mediated sugar translocation across the brush border membrane of hamster small intestine. Biochim. Biophys. Acta 193, 168-178 (1969).
4. CRANE, R. K.: Absorption of sugars. In Handbook of Physiology, Section 6: Alimentary Canal, Vol. III, C. F. Code (editor), American Physiological Society, Washington, D. C., 1968, p. 1323-1351.
5. CRANE, R. K.: Organisation der digestiv-absorptiven Funktion an der Membran des Bürstensaumes, In: Biochemische und klinische Aspekte der Zuckerresorption, K. Rommel (editor), F. K. Schattauer Verlag, Stuttgart New York, 1970, p. 75-83.
6. CRANE, R. K.: Structural and functional organization of an epithelial cell brush border, In: Intracellular transport, edited by K. B. Warren, New York, Academic Press, 1967, Vol. V, p. 71-103.
7. CRANE, R. K.: A perspective of digestive-absorptive function. American J. of Clin. Nutrition 22, 242-248 (1969).
8. CRANE, R. K., MALATHI, P., CASPARY, W. F., RAMASWAMY, K.: Evidence for a second glucose transport system in hamster small intestine specific for glucose released by brush border digestive enzymes. Federation Proceedings 29, 595 (abstract) (1970).
9. GARDNER, J. D., BROWN, M. S., LASTER, L.: The columnar epithelial cell of the small intestine: digestion and transport. New England J. of Medicine 283, 1196-1202 (1970).
10. GRAY, G. M., INGELFINGER, F. J.: Intestinal absorption of sucrose in man: interrelation of hydrolysis and monosaccharide product absorption. J. Clin. Investigation 45, 388-398 (1965).
11. HAMILTON, J. D., McMICHAEL, H. B.: Role of the microvillus in the absorption of disaccharides. The Lancet II, 154-155 (1968).
12. LYON, I., CRANE, R. K.: Studies on transmural potential in vitro in relation to intestinal absorption. 1. Apparent Michaelis constants for $Na^+$-dependent sugar transport. Biochim. Biophys. Acta 112, 278-286 (1966).
13. MILLER, D., CRANE, R. K.: Digestive function of the epithelium of the small intestine: 1. An intracellular locus of disaccharide and sugar phosphate hydrolysis. Biochim. Biophys. Acta 52, 281-293 (1961).
14. PARSONS, D. S., PRITCHARD, J. S.: Relationship between disaccharide hydrolysis and sugar transport in amphibian small intestine. J. Physiology 212, 299-319 (1971).
15. PARSONS, D. S.: Black box models of intestinal mucosal cellular function. In: Intestinal transport of electrolytes, amino acids and sugars. W. McD. Armstrong (editor), Charles Thomas Publ., Springfield, Illinois, 1971, p. 24-51.

16. RAMASWAMY, K., MALATHI, P., CRANE, R. K.: The direct transport of
glucose from disaccharides by in vitro hamster small intestine. Federation
Proceedings 30, 538 (abstract) (1971).
17. SCHULTZ, S. G., ZALUSKY, R.: Ion transport in isolated rabbit ileum. II. The
Interaction between active sodium and sugar transport. J. General Physiology 47, 1043-1059 (1964).
18. SEMENZA, G., TOSI, R., VALLOTON-DELACHAUX, M. C., MÜHLHAUPT,
E.: Sodium activation of human intestinal sucrase and its possible significance in the enzymatic organization of brush borders. Biochim. Biophys.
Acta 89, 109-116 (1964).

# The Na-Independent Transport of Sugar in Renal Tubular Cells

Arnost Kleinzeller
Department of Physiology, University of Pennsylvania School of Medicine,
Philadelphia, Pennsylvania 19104, USA

It has been shown previously (1, 2, 3, 4) that in cells of renal cortex the active accumulation of 2-deoxyhexoses proceeds by a transport mechanism which is independent of $Na^+$, and consequently, is also insensitive to ouabain. At first, it was thought that the Na-dependent active transport of many hexoses shared with the Na-independent transport system a common pathway (1, 4, 5). This view was based primarily on a set of observations showing mutual competition for sites between sugars requiring $Na^+$ for their transport and 2-deoxyhexoses (1, 2, 4); such competition was also observed (5) in the absence of $Na^+$, i. e. the active accumulation of 2-deoxygalactose was inhibited, e. g. by D-galactose, and there was reasonable agreement (2) between the $K_m$ for galactose accumulation and the $K_I$ for the galactose inhibition of 2-deoxygalactose transport, as would be expected for a shared carrier.

Subsequently, some unexpected results were obtained and the view of a common carrier was weakened in the light of the following observations:

Firstly, major discrepancies were noted (2) for some combinations of sugars between the independently determined transport $K_m$ and the $K_i$ for the competitive inhibition.

Secondly, variation of saline pH between 6. 2 and 8. 2 affected only little the (apparently) strictly Na-dependent transport of $\alpha$-methyl-D-glucoside whereas the Na-independent active accumulation of 2-deoxy-D-galactose was markedly depressed by increasing pH (ref. 6). Curiously, the Na-independent transport of 2-deoxy-glucose was actually considerably stimulated by increasing pH. Such findings were considered to be incompatible with a carrier common to $\alpha$-methyl-glucoside and 2-deoxygalactose, although the transport of both sugars was competitively inhibited by some sugars, e. g. D-galactose. Moreover, the differences in the response of the active accumulation of both 2-deoxyhexoses to variations of pH also indicated differences in the pathways of the Na-independent transport systems for these sugars.

Moreover, evidence has been obtained (3, 4) to show that D-galactose, previously believed (7) to be actively accumulated in renal cells only in the presence of $Na^+$, was to a considerable degree transported by a Na-independent mechanism.

The purpose of the present report is to:

1. Summarize the known characteristics of the Na-independent (ouabain-insensitive) transport mechanism for some sugars in renal tubular cells, and

2. Report further evidence to show that several transport pathways for sugars are operative in renal tubular cells.

These findings may provide a rationale for some of the  discrepancies pointed out above.


## Experimental methods

Unless otherwise stated, details of the methods used are reported in ref. 1, 2, 3.


## Results

As opposed to the Na-dependent (and ouabain-sensitive) transport system for sugars such as $\alpha$-methyl-D-glucoside, the active transport of 2-deoxy-D-hexoses proceeds in the absence of external $Na^+$ or in Na saline in the presence of ouabain concentrations sufficient to inhibit the operation of the Na pump (1, 2, 3, 4). No specificity for the cation replacing $Na^+$ was found (1, 3).

The Na-independent transport of 2-deoxyhexoses was also found (3) to be independent of external $K^+$. The rate of 2-deoxygalactose transport was, within the limits of experimental error, identical in standard saline (6.5 mM $K^+$) and in K-free medium, whereas for the Na-dependent accumulation of $\alpha$-methyl-D-glucoside a stimulation by $K^+$ was found, the $K^+$ requirement being saturated at $\left[K^+\right]_o$ at about 1 mM.

Both the Na-dependent and Na-independent transport systems for sugars share a requirement for $Ca^{2+}$ which is saturated at 0.5 mM $\left[Ca^{2+}\right]_o$. No evidence was obtained suggesting that the uphill transport of 2-deoxysugars might be coupled to the downhill flux of another solute; in particular, no indication for the coupling between the transport of 2-deoxyhexoses and bulk cations or anions was found (3). The available evidence thus suggests that the Na-independent active sugar transport in renal cells is directly coupled to a source of metabolic energy.

Concerning the structural requirements of the Na-independent sugar transport, it has been reported previously that a portion of the active accumulation of D-galactose in renal cells proceeded also in the absence of $Na^+$, or in the presence of 0,5 mM ouabain (ref. 3, 4). Furthermore, two separate pathways of galactose transport were indicated by the observation (6) that the fraction of the Na-dependent transport of this sugar increased with increasing pH, whereas at pH 6.2 practically all the active transport appeared to be independent of $Na^+$, at pH 8.2 the steady-state level of D-galactose in Na-free media (Li, tris- or choline salines) was only some 25% of that found in Na salines. The Na-independent transport of D-galactose reached a steady state within 60 min. of incubation and the accumulation ratio of this sugar ($S_i/S_o$) showed the same dependence on $S_o$, the external sugar concentration (Fig. 1), as reported previously for the transport in Na salines. These results emphasize the conclusion that for the Na-independent transport of sugars the structural requirement is not restricted to the absence of a hydroxyl group on $C_2$. Even for the apparently strictly Na-dependent transport of $\alpha$-methylglucoside it has been shown that, by decreasing the saline pH to 6.2, some active accumulation of this sugar could be observed in the absence of $Na^+$, and a Li gradient appeared to be involved in this transport.

In the light of the above observations it is obvious that no conclusion can be drawn as to the involvement of a single carrier in both the Na-dependent and Na-independent transport of sugars on the basis of a competitive (or partially competitive,

Fig. 1. Effect of the substrate concentration on the Na-independent accumulation of D-galactose in slices of rabbit kidney cortex. Slices were incubated aerobically $(O_2)$ at $25^O$ for 60 min in Li salines containing varying concentrations of $1\text{-}^{14}C$-D-galactose. Results are expressed as the accumulation ratio $S_i/S_O$, S denoting the apparent sugar concentration in the intracellular (subser. i) and extracellular (o) compartments. Each value of $S_i/S_O$ is the mean of 3 analyses

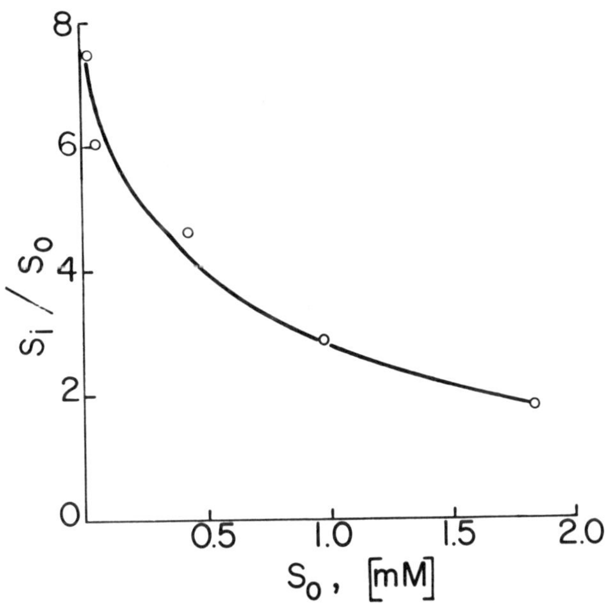


ref. 4), inhibition of the cellular accumulation of 2-deoxyhexoses by galactose, even if this takes place in the absence of $Na^+$.

We have reported earlier that phlorrhizin was a very potent inhibitor of the transport of $\alpha$-methyl-D-glucoside whereas relatively high concentrations of the inhibitor were required to demonstrate an inhibitory action on the accumulation of 2-deoxy-hexoses (1). The effect of phlorrhizin and phloretin on the transport of some sugars has now been investigated in more detail. The results of this study are summarized in Table 1.

| Sugar | Phlorrhizin $K_i$ | Phloretin $K_i$ |
|---|---|---|
| Methyl-D-glucoside | $7 \times 10^{-6}$ M | $2 \times 10^{-4}$ M |
| 2-Deoxy-D-glucose | $6 \times 10^{-4}$ M | $2 \times 10^{-4}$ M |
| 2-Deoxy-D-galactose | $9 \times 10^{-4}$ M | $5 \times 10^{-4}$ M |

Table 1. Inhibition of sugar transport in slices of rabbit kidney cortex by phlorrhizin and phloretin. Slices were preincubated aerobically $(O_2)$ in standard saline of the Krebs-Ringer type at $25^O$ for 45 min. in order to reach a steady-state level of tissue components, then incubated 20 min. in salines containing 0.5 and 1 mM of the respective sugars and varying concentrations of the inhibitor. Values of $K_i$ were computed using the Hunter-Downs plot. All values are the means of at least 3 experiments

It will be seen that the $K_i$ for the phlorrhizin inhibition of $\alpha$-methyl-D-glucoside transport was of the order of $1 \times 10^{-5}$ M, and phloretin was about 50 fold less potent as an inhibitor. On the other hand, the transport of both 2-deoxy-D-hexoses was characterized by the opposite behavior with regard to the inhibition by both substances, i. e. phloretin was somewhat more potent as inhibitor than phlorrhizin, the $K_i$ values for both substances being markedly higher than $10^{-4}$ M. It should be noted that the $K_i$ for the phlorrhizin inhibition of D-galactose transport in the presence of $Na^+$ was found to be $6 \times 10^{-5}$ M, i. e. considerably more than that for $\alpha$ - methyl-glucoside; in the absence of $Na^+$, the $K_i$ was $4 \times 10^{-4}$ M, i. e. of the same order as that found with 2-deoxysugars. This result indicates that the $K_i$ value obtained in Na salines was a composite one, representing the inhibition of both the Na-dependent and Na-independent transport pathways for galactose.

The reported differences in the $K_i$ to phlorrhizin and phloretin are not compatible with the concept of one transport mechanism being responsible for both the Na-dependent and Na-independent sugar transport (one carrier): at least two transport pathways appear to be indicated, with markedly differing affinities for both phlorrhizin and phloretin.

The discrepancies between the observed transport $K_m$ and $K_i$ values of some competing sugars also indicated separate transport pathways. This was particularly borne out by recent experiments with 2-deoxy-D-galactose-$^3$H. We reported previously that 2-deoxy-galactose competitively inhibited the transport of 2-deoxy-D-glucose, with a $K_i$ of 3. 5 mM; the independently determined $K_m$ for 2-deoxygalactose was 3. 0 mM, i. e. within the limits of experimental error these values were sufficiently close for the assumption that one carrier (one pathway) was involved in the transport of both 2-deoxyhexoses. The availability of 2-deoxy-D-galactose-$^3$H allowed to carry out the reverse experiment, i. e. to test the effect of 2-deoxy-D-glucose on the transport of 2-deoxy-galactose. No inhibitory action was found even at extreme (1:20) molar ratios of both sugars. Such results give further support to the conclusion originally based on the differences in the response to variations in pH of the transport systems for 2-deoxy-D-glucose and 2-deoxy-D-galactose, that two separate transport pathways are involved, competition taking place only at one of these.

A recent set of observations may be helpful in resolving some of the reported discrepancies. The preparation of teased flounder (Pseudopleuronectes americanus) renal tubules has been used for some time in order to study processes localized at the peritubular face of the renal tubular cells; the teased tubules form little cysts, and the luminal face of the cells is not accessible to solutes in the suspending medium (8, 9, 10). Using this preparation, it was shown (11) that some sugars, e. g. $\alpha$-methyl-D-glucoside, L-glucose, 2-deoxy-D-galactose, do not enter the cells since the steady-state tissue/medium ratio corresponded essentially to the filling of the extracellular (inulin) space and the filling of this space by the sugars was not affected by interference with tissue metabolism (e. g. dinitrophenol), other sugars, or inhibitors of sugar transport such as phlorrhizin. D-galactose, which was shown to be reabsorbed from the lumen of the flounder kidney tubule did enter the tubular cells by what appeared to be facilitated diffusion, since this transport was never observed to be up-hill (Fig. 2), was inhibited by phlorrhizin and other gars (e. g. glucose) but was not affected by metabolic inhibitors. Finally, 2-deoxy-D-glucose was shown to be taken up by the tissue against a significant concentration gradient and this transport was inhibited by dinitrophenol (Fig. 2). Moreover, the uptake of 2-deoxy-D-glucose was found to be inhibited by phlorrhizin, phloretin and other competing sugars (e. g. glucose), but not by 2-deoxy-galactose, and showed the same responses to variations of pH as previously reported for the accumulation of this sugar in slices of rabbit kidney cortex (Fig. 3). Thus, the available data

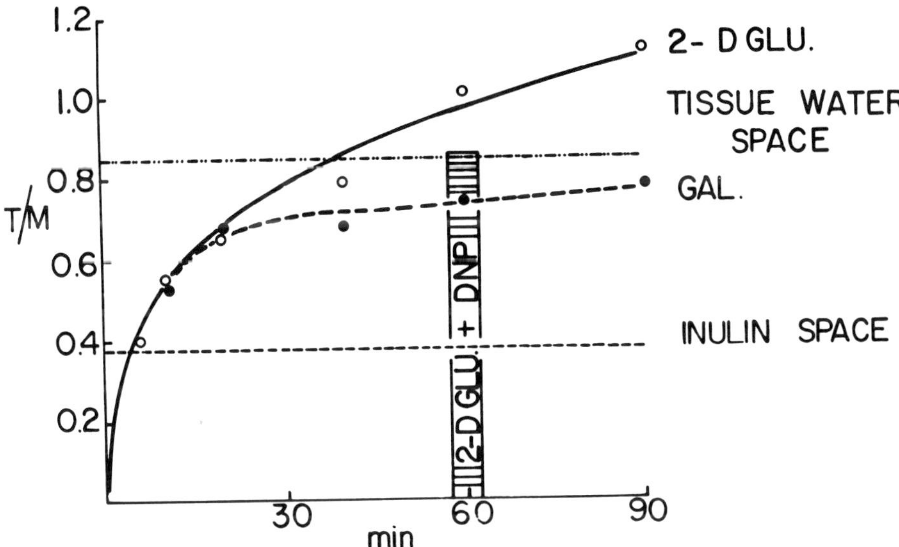



Fig. 2. Transport of D-galactose and 2-deoxy-D-glucose in teased tubules of flounder kidney. Tissue was incubated aerobically (air) at $15^0$ for 2 h in saline (ref. 9) containing 1 mM 1-$^{14}$C -D-galactose (GAL,●) or 1-$^{14}$C -2-deoxy-D-glucose (2-DGLU, o). The results are expressed as the sugar distribution ratio T/M,i. e. the amount of sugar in 1 g tissue (T) related to the amount of sugar in 1 g medium (M). Each point of the T/M is the mean of not less than 3 analyses. Column: T/M for 2-deoxyglucose found in the presence of 0. 1 mM dinitrophenol

allow the conclusion that in the flounder kidney 2-deoxyglucose is actively transported into renal tubular cells by a process localized at the peritubular face of the cells, and the transport pathway differs from that of 2-deoxy-galactose. On the other hand, the active step for the D-galactose transport in the flounder kidney tubule appears to be localized at the luminal face since during the reabsorption of this sugar a cellular accumulation of galactose took place, and no active component could be discerned at the peritubular face of the cells.

It should be noted that at least three pathways of sugar transport in renal tubular cells of dogs have been postulated (12) on the basis of experiments in vivo. As to the multiplicity of transport pathways, sugars do not appear to differ from amino acids where several transport systems with overlapping specificities are operative (13).

Parallel pathways of sugar entry into cells of functional polarity could greatly affect the observed kinetics when using a preparation such as a slice. A competitive inhibition would be observed if a carrier at the luminal brush border region of the cell would be faced with sugars S and R only from the luminal side, whereas counterflow would accelerate the entry of S into the cells if R entered at the peritubular face and competed for the carrier at the luminal face. This second mechanism has been suggested in order to explain the secretion of L-glucose into the renal tubular lumen (15). With several carriers involved at the luminal and peritubular faces of the cells, no prediction as to kinetics can be made when using a preparation with such complex geometry as a slice of kidney cortex. Obviously,

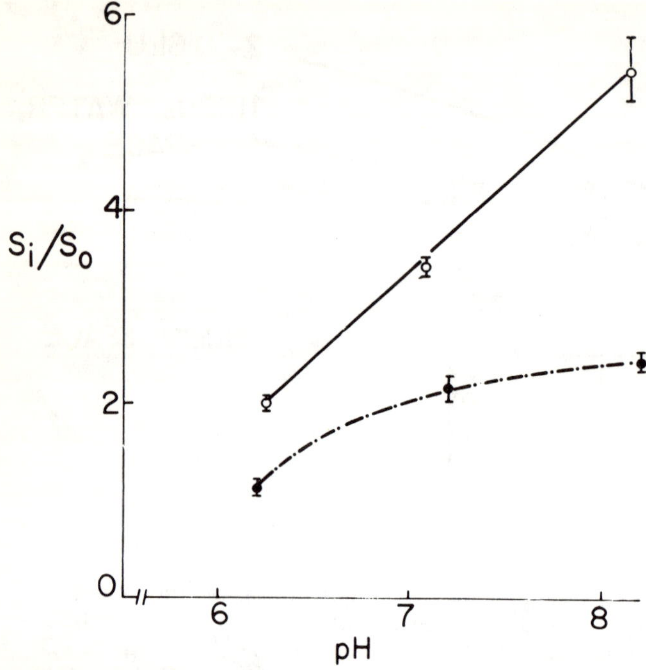

Fig. 3. Effect of pH on the $S_i/S_o$ for 2-deoxy-D-glucose in slices of rabbit kidney cortex and flounder tubules. Tissue preparations were incubated in Na salines of varying pH containing 1 mM 1-$^{14}$C-2-deoxy-D-glucose. Slices of rabbit kidney cortex (o): Incubation 60 min at 25$^o$; teased flounder kidney tubules (•): incubation 2 h at 15$^o$. Each point is the mean of at least 5 analyses, $\pm$ SE

in order to specify the properties of the carriers further, a preparation would be required which would allow to distinguish between the transport systems at both faces of the cells.

In conclusion, the results summarized here show that in renal tubular cells:

1. The Na-independent transport of sugars is not limited to 2-deoxyhexoses;

2. Under some experimental conditions, the presence of $Na^+$ does not appear to be mandatory even for the apparently strictly Na-dependent transport of $\alpha$-methyl-glucoside;

3. At least three separate transport pathways for sugars with some overlapping specificities are present.

Independence of $Na^+$ has been demonstrated also for the active transport of various sugars in the chlorioid plexus (16). Furthermore, for the intestinal mucosal cells it has been shown (17) that the active accumulation of D-xylose was practically identical in Na and Li salines; also, fructose was found to enter the cells by a mechanism which was not inhibited by ouabain or the absence of $Na^+$, although it was left open whether this transport was of the equilibrating or active type (18). Thus, a Na-independent mechanism for sugar transport is not restricted to renal tubular cells only. It would be of interest to know whether the characteristics for the Na-independent sugar transport in renal cells, pointed out above, hold also for other cell types.

This work was supported by the National Institute of Arthritis and Metabolic Diseases (Grant No. AM-12619) and a grant-in-aid from the Smith, Kline, and French Laboratories. The able assistance of Miss. E. McAvoy with some of the experiments is appreciated.

## References

1. KLEINZELLER, A., KOLINSKA, J., BENES, I.: Biochem. J. <u>104</u>, 852 (1967).
2. KLEINZELLER, A.: Biochim. Biophys. Acta <u>211</u>, 264 (1970).
3. KLEINZELLER, A.: Biochim. Biophys. Acta <u>211</u>, 277 (1970).
4. KOLINSKA, J.: Biochim. Biophys. Acta <u>219</u>, 200 (1970).
5. KLEINZELLER, A.: in W. McD. Armstrong and A. S. Nunn, Intestinal transport of electrolytes, amino acids and sugars, Ch. C. Thomas, Springfield, Ill., 1970, p. 232.
6. KLEINZELLER, A., AUSIELLO, D. A., ALMENDARES, J. A., DAVIS, A. H.: Biochim. Biophys. Acta <u>211</u>, 293 (1970).
7. KLEINZELLER, A., KOTYK, A.: Biochim. Biophys. Acta <u>54</u>, 367 (1961).
8. FORSTER, R. P.: Science <u>108</u>, 65 (1948).
9. KINTER, W. B.: Am. J. Physiol <u>211</u>, 1152 (1966).
10. MAACK, T., KINTER, W. B.: Am. J. Physiol. <u>216</u>, 1034 (1969).
11. KLEINZELLER, A., HOGBEN, L.: The Bulletin, Mount Desert Island Biol. Lab. 10, 34 (1970).
12. SILVERMAN, M., AGANON, M. A., CHINARD, F. P.: Am. J. Physiol. <u>218</u>, 743 (1970).
13. CHRISTENSEN, H. N.: This symposium.
14. BAUMANN, K., HUANG, K. C.: Pflügers Arch. <u>305</u>, 155 (1969).
15. CSAKY, T. Z., RIGOR, B. M.: in A. Lajtha and D. H. Ford, Progr. in Brain Research <u>29</u>, 147 (1968).
16. FAUST, R. G., HOLLIFIELD, J. W., LEADBETTER, M. G.: Nature <u>215</u>, 1297 (1967).
17. SCHULTZ, S. G., STRECKER, C. K.: Biochim. Biophys. Acta <u>211</u>, 586 (1970).

# Sodium-Dependent Accumulation of Sugars by Isolated Intestinal Cells. Evidence for a Mechanism not Dependent on the Na$^+$ Gradient

George A. Kimmich

Dept. of Rad. Biology and Biophysics, University of Rochester, Rochester, N. Y. 14620, USA

Having just learned that I would have an opportunity to make a formal statement when I walked in the room this morning I'm not quite sure even yet what I will say. I think it is perhaps best, however, if I begin with the data which led to my digression from what many of you consider the path of the righteous. In view of the religious overtones of this meeting which have been pointed out repeatedly this morning I'm sure that some of the things that I will say sound like heresy. I hasten to point out that there is no attempt on my part to be intentionally heretical. Therefore if I'm to be charged at the end of this talk, I trust the charge will not be mechanistic murder as there was no premeditation and then an attempt to obtain the data. Instead, we had the data in hand first and our ideas evolved from it. A few moments ago we heard a "confession" from Dr. Heinz. Perhaps the next hour should be considered as the morning's attempt at your "conversion".

The sodium gradient hypothesis for the mechanism of sodium-dependent transport systems envisions active accumulation of certain sugars and amino acids supported by energy derived from the trans-membrane cellular sodium gradient (1). A mobile membrane carrier is postulated with binding sites for both the organicmolecule to be transported and for sodium ion. In the high sodium environment characteristic of extracellular fluids, a sodium-carrier-sugar ternary complex is thought to be readily formed simply by mass-action considerations. Following translocation to the low sodium environment of the inner membrane surface, the complex dissociates to free carrier, sugar and sodium ion. The dissociated sodium ion is then available for active extrusion to the cell's environment by the so-called sodium pump. The cycle is repeated until the ternary complex forms with equal ease at both membrane surfaces. Before this point is reached, sugar concentration intracellularly must be higher than in the medium, due to the assymetry in Na$^+$ distribution maintained by the sodium pump. Some evidence indicates the carrier has a higher affinity for sugar when Na$^+$ is bound than when in a Na$^+$-free state. This would contribute to the magnitude of the sugar gradient able to be produced.

Two concepts implicit in the Na$^+$-gradient hypothesis offer a valuable approach for critical evaluation of the model. First, the model suggests that there is no direct input of metabolic energy at the locus of the sugar carrier. Second the direction in which the generated sugar gradient is established should be determined simply by the sense of the sodium ion assymetry. Therefore if the sodium gradient is reversed from the normal direction, then there should be an active extrusion of sugar from the cells! We decided to try and evaluate these concepts with the use of intestinal epithelial cells isolated from chicken small intestine by methods we reported last year (2). As we demonstrated at that time, such cells can be easily handled in suspension by micro-pipet and readily subjected to medium changes and rapid sampling techniques. They retain their functional capability as indicated by their metabolic activity and ability to accumulate sugars and amino acids against considerable gradients and in a manner characteristic of such cells before isolation

Fig. 1. Kinetics of $^{14}$C-galactose accumulation by isolated intestinal epithelial cells from chick small intestine. Data from Kimmich, 1970 (2)

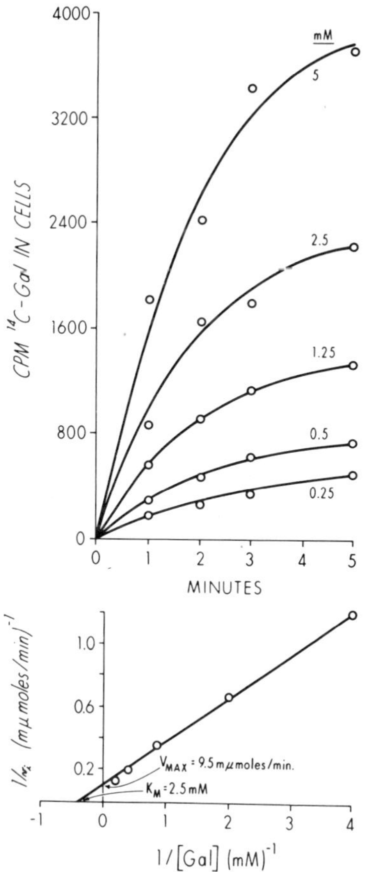



from their underlying muscle tissue. They offer distinct advantages over the various preparations of intact intestinal tissue more commonly used in that intracellular volume and hence concentration gradients can be much more precisely defined. Millipore filtration techniques allow rapid separation of cells from media and hence the opportunity for kinetic studies. In contrast, everted sacs or tissue rings are not easily manipulated and frequently samples taken at only one point in time are used to determine transport rates and concentration gradients. Often the intervals allowed are too long to provide values for initial transport velocities and potentially important information is lost.

Figure 1 simply illustrates the fact that the isolated cells do indeed accumulate sugar at a rate and to an extent which is proportional to the sugar concentration in the bathing medium. In this case, and in many of the following figures, accumulation of $^{14}$C-galactose is shown, but entirely analogous results have also been obtained with 3-0-methyl-glucose. The rate of accumulation is approximately linear for the first two minutes of incubation, and this rate was used as an estimate of initial entry velocity to construct the lineweaver-Burk plot. The indicated $K_T$ of 2. 5 mM agrees well with reported literature values obtained with more intact tissue preparations as does the $V_M$ of 250 mmoles/1 cell $H_2O$/hour.

Figure 2 indicates that the accumulation observed is active. A period of entry was allowed after which an inhibitor of cellular energy transfer was introduced. The loss of previously accumulated material following the addition of DNP demonstrates the fact that galactose had been actively accumulated prior to the interference of uptake by the inhibitor. The loss, which represents diffusional flux of galactose down its concentration gradient, proceeds until cell sugar concentration is the same as would have been achieved by the cells if uncoupler had been present from the start. The experiment also shows that phloridzin, a rather specific inhibitor of cellular sugar transport, is completely effective in preventing active sugar accumulation by the cells and allows sugar to enter only to the point where a distribution ratio of unity is attained. Ouabain is also a rather effective inhibitor of the

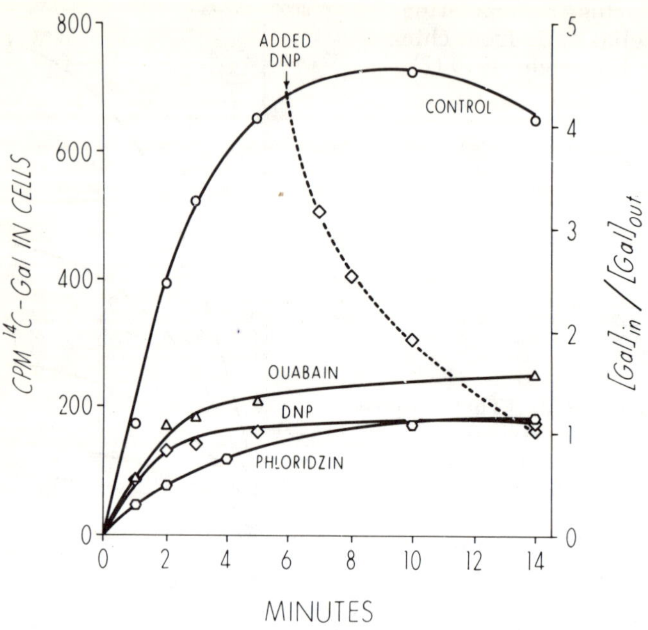



Fig. 2. Effect of ouabain, dinitrophenol (DNP), and phloridzin on accumulation of $^{14}$C-galactose by isolated intestinal epithelial cells. Data from Kimmich, 1970 (2)

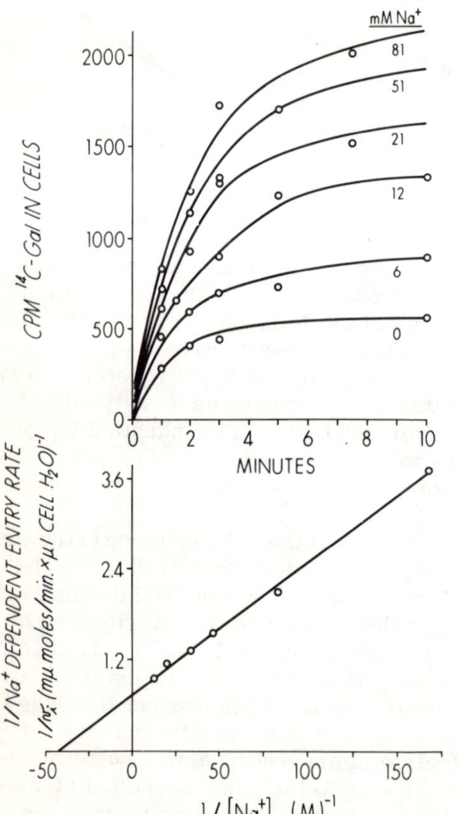



sugar accumulation process, as shown. The concentrations of inhibitors used in this experiment were 200 $\mu$M for DNP and phloridzin, and 125 $\mu$M for ouabain. In all of the following experiments where inhibibitors were used, the same concentrations were employed. In those cases where oligomycin was used it was added to the final concentration of 5 $\mu$M/ml.

The data in Figure 3 show that the galactose transport exhibited by the cells is sodium-dependent, as it is in intact intestinal tissue. The process exhibits a $K_T^{Na}$ of about 20 mM, as has been reported for other systems. Only the $Na^+$-dependent initial entry rates were used in constructing the reciprocal plot.

Fig. 3. Dependence of galactose accumulation by the isolated intestinal cells on sodium ion. From Kimmich, 1970 (3)

Fig. 4. Fluxes of $^{22}$Na$^+$ and
$^{14}$C-galactose in intestinal
cells on which a reversed
sodium gradient has been imposed. From Kimmich, 1970
(3)

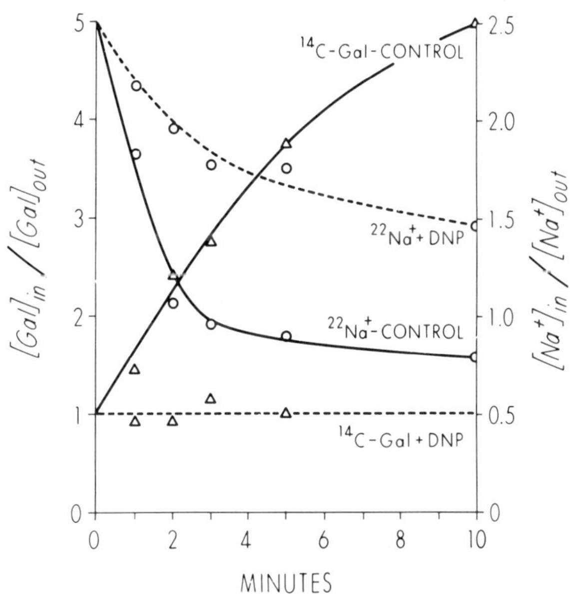



Figure 4 shows the fluxes of $^{22}$Na and $^{14}$C-galactose which occur when cells which
have been pre-loaded with sodium and galactose are placed in a low sodium environment. Two populations of cells from the same preparation were suspended in
a medium containing 80 mM sodium and 1.25 mM galactose. In one case the medium
included $^{22}$Na and in the other $^{14}$C-galactose. The pre-incubation was carried out
for 15 minutes at $0^{\circ}$ to allow the cells to at least partially come toward equilibrium
with respect to the distribution of sodium and galactose between medium and cell
water. At time 0 the cells were introduced to a mannitol medium at $37^{\circ}$: the dilution
of the cell suspension was such that extracellular (Na$^+$) became 20 mM. The dilution contained 1.25 mM non-radioactive galactose for those cells preincubated with
$^{22}$Na and 1.25 mM $^{14}$C-galactose for the cells pre-incubated with $^{14}$C-galactose.
In the latter case the specific activity of $^{14}$C-galactose was the same in both the
pre-incubation and experimental phases so that any flux of $^{14}$C which occurs represents a net flux of galactose. Samples of the cell population were taken at the
indicated intervals.

Cells pre-incubated with $^{22}$Na lose $^{22}$Na to their incubation medium even when incubated with a high concentration of DNP or ouabain to prohibit active sodium extrusion. This represents therefore only a diffusional or passive loss of cell sodium
and provides evidence that the cells have indeed been loaded to an intracellular
sodium concentration greater than 20 mM. This passive loss of Na$^+$ continues until a steady state is attained, indicating the point at which the activity of intracellular
sodium matches that in the bathing medium. In the absence of inhibitor, Na$^+$ efflux
also occurs, but at a more rapid rate and proceeding to a greater extent. This is
expected due to the activity of the sodium pump allowed in the absence of inhibitor.
Note that in this case approximately two minutes are required before cellular sodium
activity falls to the level of that in the medium, i.e., the level indicated by the steady state achieved by the inhibited cells. Thus, in the control cells, a reversed sodium
gradient is maintained for an interval approaching two minutes. The sodium gradient
hypothesis predicts an active extrusion of sugar should occur during this interval.
Instead, the flux of $^{14}$C-galactose indicates further entry of sugar against a con-

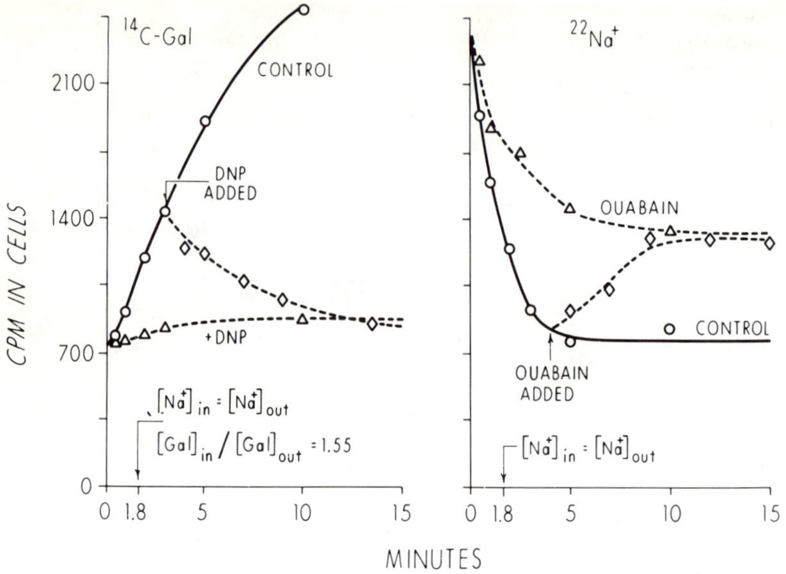



Fig. 5. Fluxes of $^{22}$Na$^+$ and $^{14}$C-galactose in isolated intestinal cells after initially imposing a reversed sodium gradient. Effect of DNP and ouabain on sodium and sugar fluxes after a period of active transport has been allowed, is also shown. From Kimmich, 1970 (3)

centration gradient. At the end of two minutes, sugar concentration in the cell water is twice that in the medium. The fact that no further entry of sugar occurs in the presence of DNP indicates that the cells had equilibrated with respect to galactose during the pre-incubation phase of the experiment. Note that the entry rate is as great during the period when the sodium gradient is reversed as at any subsequent interval. These facts are difficult to reconcile with the concept of a sugar transport mechanism which derives its energy from an inwardly directed sodium gradient.

The data illustrated in Figure 5 show the results of a similar experiment, but in this case an inhibitor was added after a brief interval of unimpeded flux had been allowed. The fact that the cellular galactose content returns to the same level as when inhibitor was included from the start indicates the sugar had indeed been accumulated actively (i. e., against a concentration gradient) during the early minutes of incubation. The $^{22}$Na flux data shows that in this instance 1.8 minutes is required for the imposed reversed Na$^+$ gradient to dissipate. At this point in time the galactose distribution ratio had already reached 1.55.

Figure 6 presents perhaps the most compelling argument of all in favor of an energy input for sugar accumulation which is divorced from the cellular sodium gradient. Here the cells were pre-incubated in 2.5 mM $^{14}$C-3-OMG and incubated at 1.25 mM $^{14}$C-3-OMG. Sodium concentrations were the same as in the previous two experiment. Loss of sugar from the cells would be predicted due to passive diffusional loss and due to the reversed sodium gradient imposed if the concepts of the sodium gradient model are correct. The observed additional 3-OMG in flux which commences immediately suggests that an energy input independent of the sodium gradient must operate. This input must be significant enough to more than compensate for the diffusion-

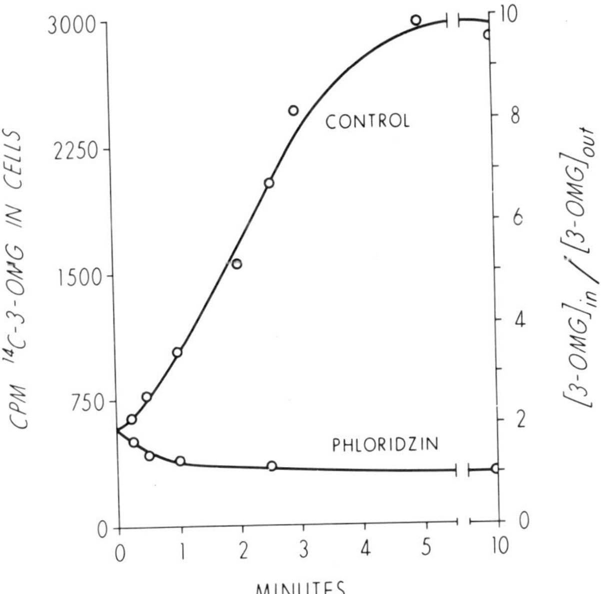



Fig. 6. Accumulation of $^{14}C$-3-OMG when outwardly directed gradients of both sodium and 3-OMG are initially imposed. Data taken from Kimmich, 1970 (3)

al loss expected from the initial outward oriented sugar gradient. The diffusional loss is demonstrable by the outward flux which occurs in the presence of phloridzin and at $0^\circ$. Note the efflux proceeds until exactly half the initial cellular 3-OMG concentration is reached, as expected.

Another aspect of the sodium gradient hypothesis provides for a possible role for the cellular potassium gradient. This idea originated from the fact that elevated $K^+$ concentrations inhibit sodium dependent transport systems in a manner which indicates competition of $K^+$ for the $Na^+$ site. Also a loss of cellular $K^+$ from intestinal tissue has been demonstrated during periods of rapid sugar or amino acid transport (6). It has been suggested that these facts may indicate that $K^+$ can fill the $Na^+$ site on the sugar carrier and produce a form which has a very low affinity for sugar (7). This would tend to occur in the high $K^+$ environment at the inner membrane surface and cause discharge of entering sugar molecules more readily. The $K^+$-carrier binary complex might then migrate to the outer surface and dissociate, producing the observed loss of cellular $K^+$. Such a mechanism implies that both monovalent ion gradients could provide energy to drive active sugar entry. It also implies a certain mobility of $K^+$ out of the cell on the sugar carrier. Several observations suggest that this carrier-mediated efflux must be rather substantial in magnitude. Namely, it has thus far been difficult to demonstrate a net increase in intestinal epithelial cell sodium during periods of active sugar or amino acid transport in normally energized tissue preparations (i. e., untreated with metabolic inhibitors) (6). This is true despite the fact that increases in uni-directional $Na^+$ influx are easily demonstrable (8), and it indicates that the active extrusion mechanism for $Na^+$ must be keeping pace with the additional $Na^+$ being delivered to the cell on the sugar carrier. Sodium extrusion is, of course, partially coupled to $K^+$ accumulation, and for these reasons one might predict increases in cell $K^+$ during sugar transport. The observed decrease must signify a significant loss of $K^+$ occurs through the sugar carrier. It is possible that agents such as phloridzin which block the function of the sugar carrier (9), might therefore prevent losses of cell $K^+$ occurring by this route and allow the cells to generate better $K^+$ gradients than

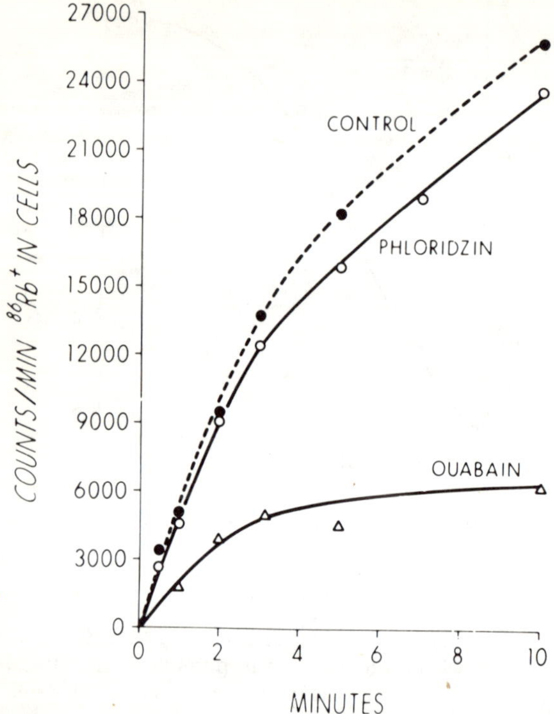



Fig. 7. Effect of phloridzin and ouabain on $^{86}$Rb uptake by isolated intestinal cells. From Randles and Kimmich, 1972 (4)

normally achieved. Figure 7 shows that this is not true for a case where $Rb^+$ accumulation is monitored. Rubidium behaves in a manner exactly analogous to $K^+$ in terms of its effects on the sugar transport capability of the isolated cells. We have never observed enhancement of either the rate or extent of rubidium or potassium uptake by phloridzin in the isolated cells. While this represents only a preliminary approach to the question at hand, we feel that the concept of $K^+$ efflux on the sugar carrier should be regarded with some caution.

On the other hand, $K^+$ exerts a powerful effect on the sugar accumulation capability of the isolated cells, just as has been reported for intact tissue preparations. Figure 8 shows the effect of several different $K^+$ concentrations on the kinetics of galactose transport by the cells. Extracellular $Na^+$ was 20 mM in this experiment. The degree of accumulation in the presence of DNP again indicates the point at which a distribution ratio of unity is achieved. As little as 4 mM $K^+$ produces a significant inhibition while 90 mM $K^+$ prevents 90% of the active galactose uptake. The question immediately arises as to whether the observed inhibition represents an effect mediated by loss of energy inherent in the $K^+$ gradient as extracellular $K^+$ is elevated. If so, the $K^+$ gradient may be providing a large

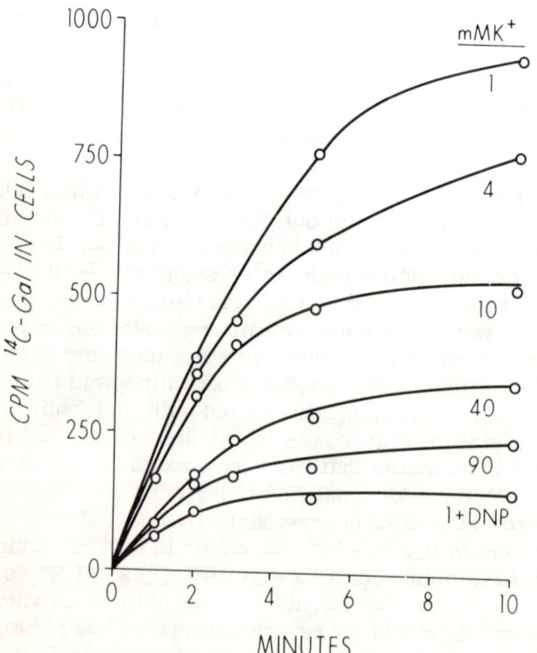



Fig. 8. The effect of elevated $K^+$ concentrations on the accumulation of $^{14}$C-galactose by isolated intestinal epithelial cells. Data taken from Kimmich, 1970 (3)

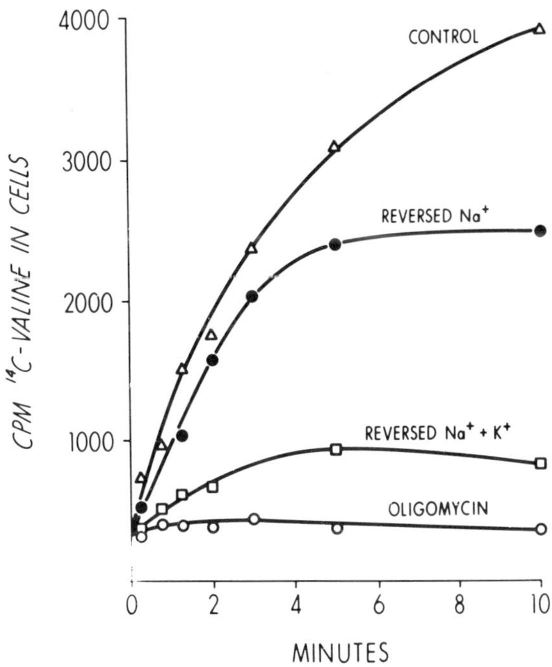



Fig. 9. Accumulation of $^{14}$C-3-OMG by isolated intestinal cells after initially imposing reversed gradients of both Na$^+$ and K$^+$. The cells were pre-equilibrated with $^{14}$C-3-OMG in the medium before starting the experimental phase. Data taken from Randles and Kimmich, 1972 (4)

portion of the energy required to support the active accumulation of sugar. The sugar entry described earlier for a situation with a reversed sodium gradient might therefore have been supplied by a normal or near-normal K$^+$ gradient. In order to evaluate this possibility we attempted to monitor sugar flux in cells pre-loaded to equilibrium with $^{14}$C-galactose immediately after imposing a reversed gradient of both Na$^+$ and K$^+$. Cells were introduced to a medium containing 20 mM Na and 60 mM K$^+$ following pre-incubation at 80 mM Na$^+$ and 6 mM K$^+$. The data are shown in Figure 9. The immediate further accumulation of galactose which occurs in the face of the unfavorable ion gradients argues against an entry system driven by normally oriented Na$^+$-K$^+$ gradients. Remember that the concepts implicit in the ion gradient hypothesis predict an active <u>extrusion</u> of galactose during the early part of the incubation period (10).

The much slower rate and smaller extent of entry that occurs in the previous experiment when both ion gradients are reversed compared to the situation when only the sodium gradient is reversed is not an indication that a large portion of the energy for sugar transport is derived from the K$^+$ assymetry. It simply reflects the fact that the K$^+$ concentration in the two situations is very different. There is no question that high concentrations of K$^+$ are very inhibitory to active intestinal sugar transport. However the basis of that inhibition does not relate to the magnitude or direction of the K$^+$ gradient existing across the cellular membrane. The proof of the last statement is best demonstrated by a situation in which cell K$^+$ levels have been manipulated, but transport of sugar monitored at a constant medium K$^+$ concentration. Such a situation is illustrated in Figure 10. One cell population was pre-incubated at 80 mM K$^+$ and the other with no K$^+$. The pre-incubation for this experiment was performed in the absence of Na$^+$ in order to facilitate loss of endogenous intracellular K$^+$ from those cells initially suspended in medium with no K$^+$. The experimental phase for both populations was performed at 20 mM K$^+$ and 60 mM Na$^+$. Thus the cells pre-incubated in a high K$^+$ medium had a much more favor-

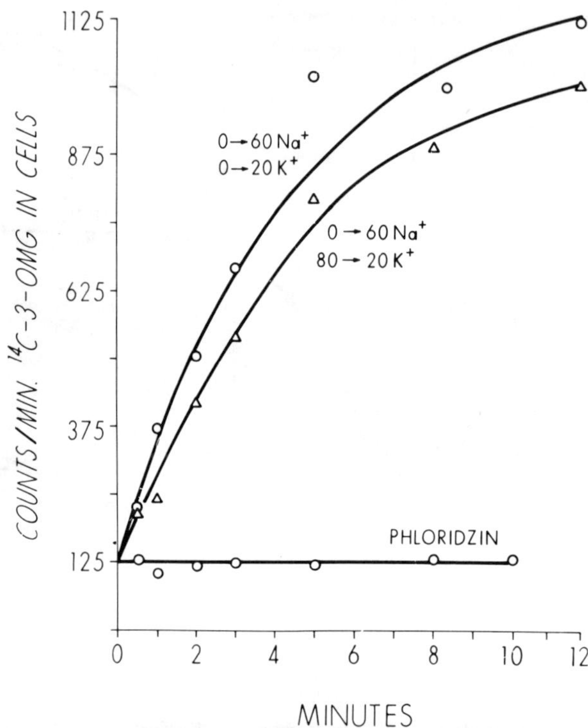



Fig. 10. Accumulation of $^{14}$C-3-OMG by isolated intestinal cells in two situations with different intracellular K$^+$ levels. Data taken from Randles and Kimmich, 1972 (4)

able normally directed K$^+$ gradient than the other cell population. The rate and extent of sugar uptake by the two groups was virtually identical. Those cells with the elevated internal K$^+$ concentration showed no better ability to concentrate 3-OMG than the others. In fact their ability was slightly impaired. Therefore, we must conclude that at a given extracellular K$^+$ concentration, sugar transport rate is not modified by the level of cellular K$^+$ concentration, i. e., the magnitude of the K$^+$ gradient.

What is the basis of the K$^+$ effect? A possible clue is provided by the data of Figure 11. It shows that the inhibitory effect of 80 mM K$^+$ on the phloridzin sensitive portion of 3-OMG uptake practically disappears if the experiment is performed with cells that have been pre-treated with dinitrophenol for 10 minutes to deplete their energy stores. These data should be compared with Figure 8 which shows that in normal cells 90 mM K$^+$ inhibits the phloridzin sensitive portion of galactose uptake be 90%. Thus the behavior of the sugar carrier toward K$^+$ ion depends on the energy state of the cell. Potassium seems effective as an inhibitor only when the cell has a capability for energy conser-

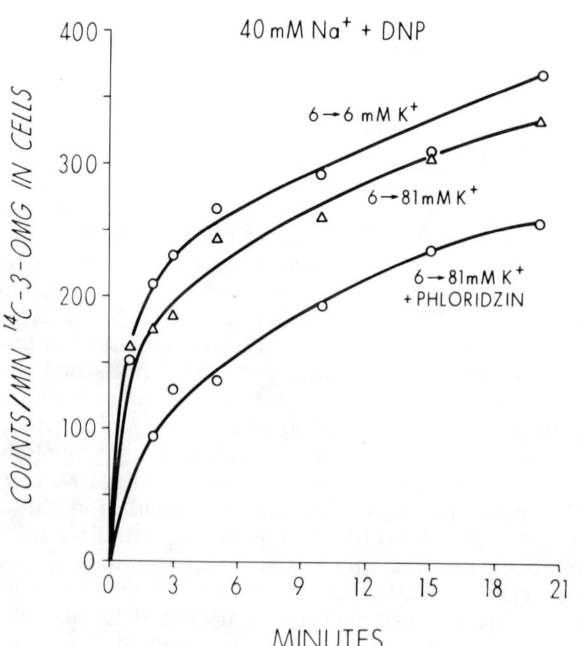



Fig. 11. The effect of K$^+$ on $^{14}$C-3-OMG accumulation in cells deenergized by a 10-minute pre-incubation with 200 $\mu$M DNP. Data from Randles and Kimmich, 1972 (4)

vation. This suggests that the properties of the sugar carrier are dependent on a direct input of metabolic energy rather than determined simply by the trans-membrane distribution of sodium and potassium.

There is one particularly serious objection to those experiments presented above in which a reversed Na$^+$ gradient was imposed. In each of these instances it has been assumed that the cellular sodium was uniformly distributed in the cell water. However, it is the sodium concentration near the inner surface of the cell membrane which will be sensed by the sugar carrier. If micro-environments exist in this locus of the cell with markedly lower sodium concentrations than occur in the bulk cell water, then the observed sugar entry might still be driven by a normal sodium gradient. This possibility appears extremely difficult to evaluate properly at first consideration. However, the characteristics of interaction between the Na-dependent transport systems for sugars and amino acids provides an answer.

A mutual inhibitory interaction between these two transport systems has been described by a number of research groups (11-13). A mechanistic basis for the interaction has been suggested which retains the basic concepts of the Na-gradient hypothesis. The essential element of this idea is that interaction occurs as a result of competition between the two transport systems for cellular energy reserves in the form of the trans-membrane sodium gradient. The co-entry of Na$^+$ which occurs when one molecular species is transported is viewed as causing a partial dissipation of the sodium gradient, leaving less energy inherent in that gradient to support transport of the other species. Inhibition is postulated because the carriers for both substrate species sense the same intra-cellular environment. Delivery of sodium to this environment by either carrier tends to decrease the dissociation of sodium from the other, and hence the second carrier. From this viewpoint it is important to realize that the effectiveness of any substrate as an inhibitor of another transport system should be correlated with its rate of entry. That is, the amount of sodium co-transported by a metabolic carrier will be related to the rate of entry of the metabolite in question. Therefore those substrates with high entry rates should be more effective in discharging the sodium gradient, and more effective as inhibitors of other transport systems deriving their energy from the same sodium gradient.

Figure 12 shows that there is a marked interaction between the transport systems for amino acids and sugars in the isolated cells. The active accumulation of valine is 75% inhibited by 10 mM 3-OMG. On the other hand, active transport of 3-OMG is only inhibited by 25% when 10 mM valine is included in the incubation medium. The greater effectiveness of sugar as an inhibitor suggests that its entry rate must be significantly higher than that for valine. Instead, initial flux measurements for the two substrates show the transport rate for valine is more than four times faster than for 3-OMG (Table I). Here we are confronted then with a situation in which valine should be discharging the sodium gradient at a rapid rate. yet it is a poor inhibitor compared to 3-OMG. Co-transport of sodium should be four-fold slower with 3-OMG, yet it is an excellent inhibitor. In short, there is no correlation between the transport rates of various species (or rate of Na$^+$ entry in co-transport) and the relative effectiveness of those species as inhibitors. Thus we find no correlation between the relative degree of discharge of the cellular sodium gradient and the degree of inhibition of transport systems supposedly deriving their energy from the sodium gradient. It is important to recognize that this argument applies whether one considers sodium concentrations in the bulk cell water or simply in micro environments existing near the membrane surfaces. If the concepts of the sodium gradient hypothesis account for the interaction between transport systems, the carriers for the two systems must be sensing a common intracellular compartment, which

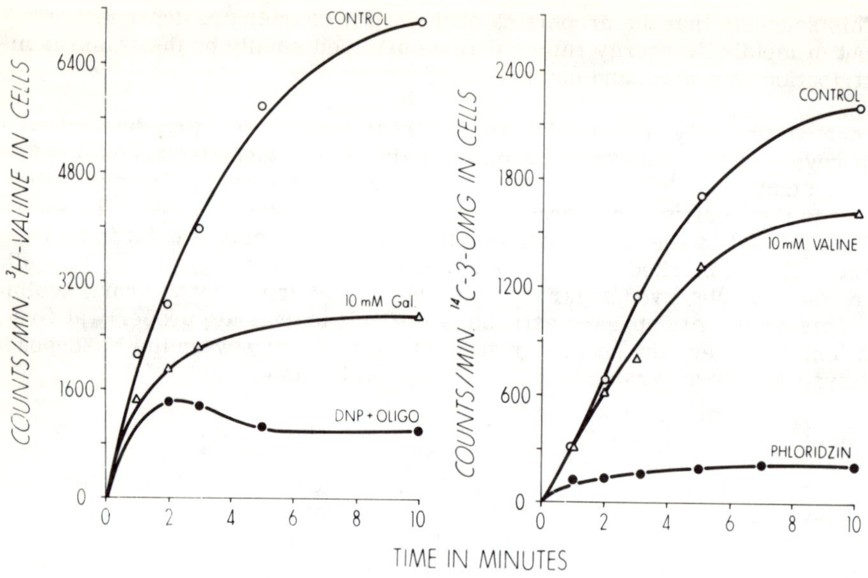



Fig. 12. Interaction between intestinal transport systems for sugars and amino acids. In the control experiments the substrate being accumulated was present at a concentration of 1 mM. Data taken from Randles et al., 1972 (5)

might be either a macro or micro compartment.

The data presented thus far represent the primary basis for my reluctance to embrace the concepts embodied in the sodium-gradient hypothesis. I feel the data presented in the previous section with regard to the interaction between transport systems is perhaps the strongest as it does not depend on an accurate determination of actual concentration gradients established (i. e., intracellular space measurements). Instead only initial flux rates need be determined in the presence or absence of the inhibitory species. These values are relatively easy to determine and have a high degree of reproducibility.

One criticism of our data that we frequently hear is the fact that we are dealing with a system in which the normal cell polarity is not maintained. The suspending medium bathes all surfaces of the cell and therefore we have no way of evaluating what fraction of the observed uptake is related to events occurring at the brush border and what part might be related to transport events at other surfaces. We feel that this argument is invalid, however, based on information which has been obtained with intact intestinal tissue in which polarity is preserved. First, in most of our work, we define _active_ transport as accumulation of sugar which is phloridzin sensitive. Phloridzin has not been shown to be inhibitory to sugar fluxes occurring at the serosal surface of intestinal epithelium. Second phloridzin sensitive transport by the cells is also $Na^+$-dependent, yet sodium-dependence has not been demonstrated as a characteristic of serosal sugar entry. Finally, the properties of sugar accumulation by the isolated cells with respect to inhibition by $K^+$ and values of kinetic constants for both sugar and amino acid transport are in good agreement with data obtained for mucosal transport systems established with intact tissue. For these reasons, we feel that the active sugar accumulation which we observe with the isolated cells reflects events occurring at the brush border or mucosal boundary.

Until this point I have not mentioned any alternative model for accommodating the characteristics of the sodium dependent transport systems. While I feel our data argues against an energy input for these systems which is derived from the cellular ion gradients, it as difficult to ignore the large number of observations which indicate characteristics common to both the electrolyte and non-electrolyte transport systems. It is tempting to speculate that these characteristics reflect more than the indirect link between the two systems envisioned by the sodium-gradient hypothesis. We have suggested one possibility for a more direct relationship which depends on the premise that the energized phosphorylated intermediates which have been described for the sodium pump (14-16) might be used to provide energy for a number of active transport systems including those for sugars and amino acids. The idea can be represented schematically as follows:

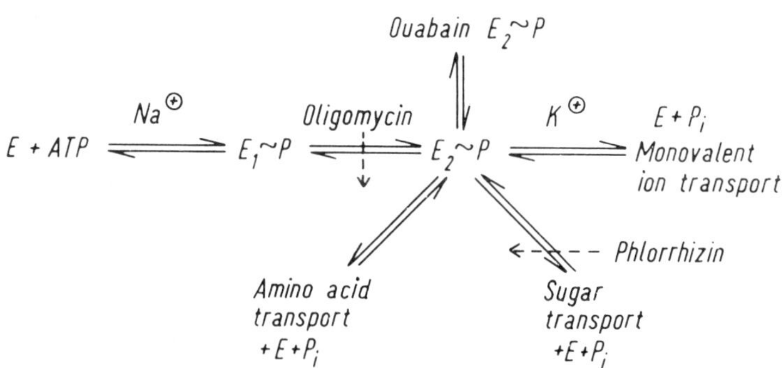


The designated terminology is that originally used by investigators interested in the mechanism of $(Na^+ + K^+)$ ATP-ase action, the enzyme thought to be an integral part of the so-called sodium pump. E represents the free enzyme, and $E_1{\sim}$ P and $E_2{\sim}$ P are phosphorylated forms of the enzyme which seem to be important in the energy transfer reactions necessary for monovalent ion transport. They may represent different conformational states of a single molecular species. Sodium is essential for the formation of the first intermediate from free enzyme and ATP, and $K^+$ is required for discharge of the second intermediate and release of energy in a manner which can be utilized to generate the usual cellular gradients of monovalent ions. Oligomycin and ouabain are thought to inhibit ion transport as indicated either by preventing interconversion between the two energized species, or by forming a complex with E or $E_2{\sim}$ P and precluding their usual energy transfer functions. It is tempting to speculate that the transport systems for sugars and amino acids might also derive energy from the pool of membrane-bound energized intermediates generated by the sodium pump as illustrated in the schematic representation. This idea provides us with a model which can accommodate much of the data which has been described regarding the characteristics of sodium-dependent transport systems. The process would be sodium-dependent because $Na^+$ is required for the reaction which generates the proper energized intermediate. Potassium ion is inhibitory because it dissipates energy from the $E_2{\sim}$ P pool, and causes it to be expended for monovalent ion transport. Ouabain and oligomycin prevent the usual energy transfer reactions for the intermediates and thus inhibit sugar and amino acid accumulation. An interaction between sugar and amino acid transport systems is predicted due to the fact that each derives its energy from the same limited pool of energized intermediate. The rates of sugar and amino acid transport would depend on the relative rates of the reactions serving to generate or dissipate $E_2{\sim}$ P, and _not_ directly on the magnitude or direction of the ion gradients. However, the direction of the ion gradients imposed might modify the activity of the non-electrolyte transport systems due to the reversibility of the ATP-ase reactions, which has been demonstrated (17). Also, a general correlation between the activity of the sodium pump and the nonelectrolyte transport system would be predicted because each derives its energy input from a common pool of membrane-bound energized intermediate. A loss of cellular $K^+$ occurs when sugars or amino acids are transported because less energy can be provided to monovalent ion transport events when $E_2 \sim P$ is being dissipated to support non-electrolyte entry. Potassium accumulation is, of course, partially linked to $Na^+$ extrusion, so that one would also predict an increase in cellular $Na^+$ when non-electrolyte is accumulated. Again this would represent diversion of energy normally channeled to $Na^+$ extrusion from the $E_2 \sim P$ pool, and its use for metabolite transport. A change in sodium extrusion at one face of the polarized epithelial cell, but not the other, would also cause the alteration in short-circuit current and transmural potential which have been observed in intact tissue preparations during periods of active sugar or amino acid transport. Finally, this model does not provide for a direct interaction of the non-electrolyte carrier with $K^+$. Instead, our model predicts $K^+$ will be inhibitory only when the proper energized intermediate can be generated from free enzyme and a supply of cellular ATP. The fact that $K^+$ loses much of its effectiveness as an inhibitor when cells are de-energized by pretreatment with DNP suggests that an indirect role for $K^+$ may have some validity. We realize that our model envisions some common components for the non-electrolyte and monovalent ion transport system in contrast to most current opinion. On the other hand, work from other laboratories indicates that as much as 50% of the $Na^+$ dependent influx of alanine across the brush border membrane of intact intestinal tissue is ouabain sensitive when the tissue is pre-incubated in normal Ringer's medium (18). This may indicate that some common components do indeed exist for the two transport systems.

## Acknowledgements

This paper was performed under contract with the U. S. Atomic Energy Commission at the University of Rochester Atomic Energy Project and has been assigned Report No. UR-49-1526.

## References


1. CRANE, R. K.: Handbook of Physiology - Alimentary Canal, 3, 1323 (1968).
2. KIMMICH, G.: Biochemistry 9, 3659 (1970).
3. KIMMICH, G.: Biochemistry 9, 3669 (1970).
4. RANDLES, J., KIMMICH, G.: submitted to Biochemica Biophysica Acta.
5. RANDLES, J., KIMMICH, G., BARRETT, G., TUCKER, A. M.: submitted to Biochimica Biophysica Acta.
6. KOOPMAN, W. G., SCHULTZ, S. G.: Biochim. Biophys. Acta 173, 338 (1969).
7. CRANE, R. K., FORSTNER, G., EICHOLZ, A.: Biochim. Biophys. Acta 109, 467 (1965).
8. CURRAN, P. F., SCHULTZ, S. G., CHEZ, R. A., FUISZ, R. E.: J. Gen. Physiol. 50, 1261 (1967).
9. CASPARY, W. F., STEVENSON, N. R., CRANE, R. K.: Biochim. Biophys. Acta 193, 168 (1969).
10. SCHULTZ, S. G., CURRAN, P. F.: Physiol. Rev. 50, 637 (1970).
11. NEWEY, H., SMYTH, D. H.: Nature 202, 400 (1964).
12. SAUNDERS, S. J., ISSELBACHER, K. J.: Biochim. Biophys. Acta 102, 397 (1966).

13. HINDMARSH, J. T., KILBY, D., WISEMAN, G.: J. Physiol. 186, 186 (1966).
14. FAHN, S., KOVAL, G. J., ALBERS, R. W.: J. Biol. Chem. 241, 1882 (1966).
15. POST, R. L., KUME, S., TOBIN, Y., ORCUTT, B., SEN, A. K.: J. Gen. Physiol. 54, 3065 (1969).
16. SEN, A. K., TOBIN, T., POST, R. L.: J. Biol. Chem. 244, 6596 (1969).
17. GARRAHAN, P. J., GLYNN, I. M.: J. Physiol. 192, 237 (1967).


## TABLE I

Comparison of Carrier Mediated Entry Ratio for Valine and 3-OMG in Isolated Intastinal Cells.

| Substrate | Entry Rate in m $\mu$ moles/min/mg protein | | |
| --- | --- | --- | --- |
| | Total | Diffusional* | Carrier-Mediated |
| 3-OMG (10 mM) | 14.5 | 1.3 | 13.2 |
| Valine (10 mM) | 61.5 | 5.4 | 56.1 |

* Diffusional entry rates were determined in the presence of agents which bind to the same carrier, and prevent carrier-mediated entry.

# The Effect of Sodium on the Transtubular Transport of D-Glucose in Rat Kidney and on the D-Glucose Binding to Isolated Brush Border Membranes

K. Baumann and R. Kinne
Max-Planck-Institut für Biophysik, Frankfurt/Main, Germany

The transtubular net flux of glucose in the proximal tubule of rat kidney has been described as an active component (6, 7) which could be blocked by adding $1 \times 10^{-4}$M phlorizin to the perfusion solution and a diffusion component (7). The calculated apparent transport parameters were:

$$V_{max\ D\text{-}gl.} \quad \text{equal to } 6 \times 10^{-10} \text{mol} \times cm^{-2} \times sec^{-1}$$

$$K_{T\ D\text{-}gl.} \quad \text{equal to } 0.6 \times 10^{-3}M$$

$$P_{D\text{-}gl.} \quad \text{equal to } 1.7 \times 10^{-5} cm \times sec^{-1}$$

These microperfusion experiments were performed under the condition of zero net flux of water and electrolytes (sodium concentration 110 mEq). Under the condition of normal net flux of water and electrolytes (sodium concentration 145 mEq), the above-mentioned apparent transport parameters were in good agreement with those calculated from the measured intraluminal glucose concentration in the second part of the proximal tubule under free flow conditions (4).

The binding studies of tritiated phlorizin (2, 3) to isolated brush border membranes (5) revealed, in a plot of the ratio of bound phlorizin (b) to free phlorizin (f) against bound phlorizin (b), at least two binding components; one with higher phlorizin affinity and the other with a lower affinity (Fig. 1). The high-affinity binding of phlorizin could be blocked in part by adding 0.04 mol/1 D-glucose. This inhibition is complete with 0.4 mol/1 D-glucose without influencing the low-affinity binding.

In these experiments the isolated brush border fragments were incubated with various concentrations of tritiated phlorizin in a buffered solution. After a 30-minute incubation period at 37°C, the samples were immediately centrifuged at high speed at room temperature for 10 minutes. The supernatant was separated carefully from the pellet and both a small sample of the supernatant and the whole pellet were analyzed for radioactivity. From the pellet counting, the amount of tritiated phlorizin (b) which was actually bound to the brush border fragments was calculated. The amount of free phlorizin (f) was determined by supernatant counting.

As shown on a Lineweaver-Burk plot (Fig. 2), this high-affinity binding was blocked in a fully competitive manner by adding D-glucose to the incubation medium. L-glucose did not block the high affinity binding as would be expected from previous

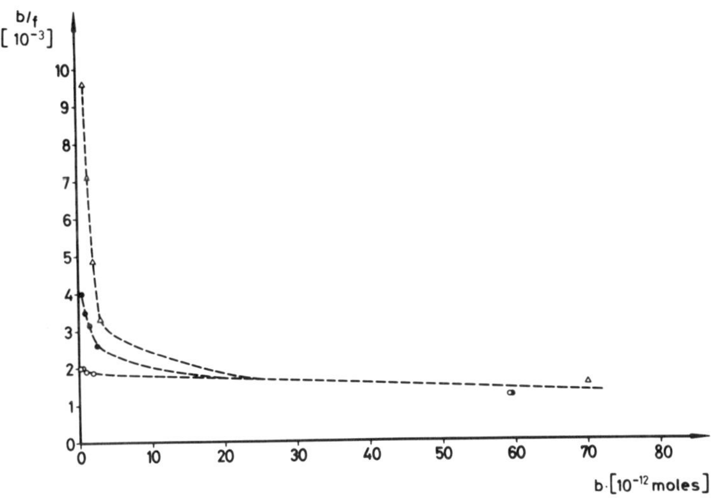

Fig. 1. Tritiated phlorizin binding to isolated brush border membranes and the effect of various D-glucose concentrations in the incubation medium.
Ordinate: the ratio of bound phlorizin to free phlorizin (b/f).
Abscissa: amount of bound phlorizin (b)
The amount of alkaline phosphatase of brush border membranes is used as a reference point. Each point represents the mean of all single data obtained at a certain concentration of phlorizin and with 150 mEq/l sodium in the medium.
Δ phlorizin binding in the absence of D-glucose
● phlorizin binding in the presence of 0.04 mol/l D-glucose
o phlorizin binding in the presence of 0.40 mol/l D-glucose
The connecting lines were drawn by simple inspection.

experiments (1). In these microperfusion experiments a secretion of L-glucose but no reabsorption was found. Fig. 3 shows the influence of sodium in the absence of glucose on the high affinity binding of tritiated phlorizin.

Both the binding of phlorizin and its inhibition by D-glucose were dependent on the sodium concentration of the incubation medium. Sodium enhanced the affinity of the binding site towards phlorizin and D-glucose while the number of binding sites remained unchanged. The calculated sodium concentration, which caused a half-maximal increase in phlorizin binding, was 13 mEq/l and for D-glucose binding, 15 mEq/l.

In preliminary experiments an attempt was made to evaluate the influence of sodium on the transtubular transport of D-glucose. Male Wistar rats were prepared in the usual manner for microperfusion experiments as described elsewhere (8). Proximal tubules were continuously perfused with test solutions containing 2 or 30 mmoles/l D-glucose $\left[\text{D-glucose-C}^{14}\ (\text{U})\right]$ . The sodium-containing perfusate had initially the following composition per liter: sodium (120 mEq), potassium (4 mEq), calcium (3 mEq), magnesium (2 mEq), chloride (109 mEq), bicarbonate (10 mEq), acetate (10 mEq), non-reabsorbable raffinose (58 or 30 mmoles), and D-glucose. Cholin (110 mEq) and lithium (10 mEq) were substituted for sodium and chloride (119 mEq) for bicarbonate in the sodium-free perfusate. The volume net flux was checked by adding inulin-methoxy-H$^3$ to the perfusion solution.

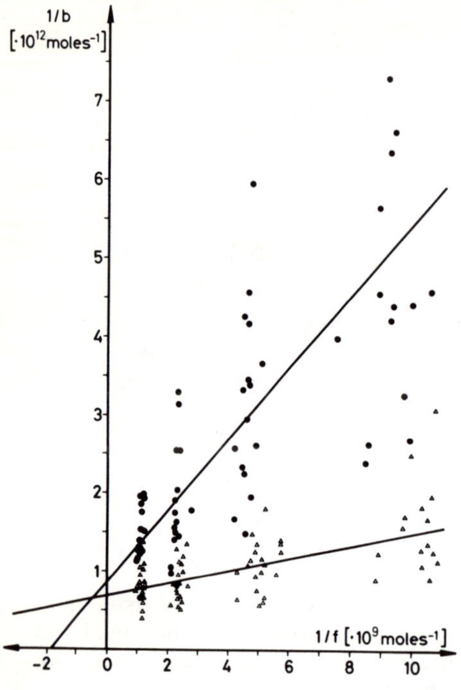

Fig. 2. Lineweaver-Burk plot of all data for phlorizin binding to the receptor site of the high affinity component. Δ incubation medium without D-glucose, ● incubation medium containing 0.04 mol/l D-glucose. The regression lines are calculated by least-square method.

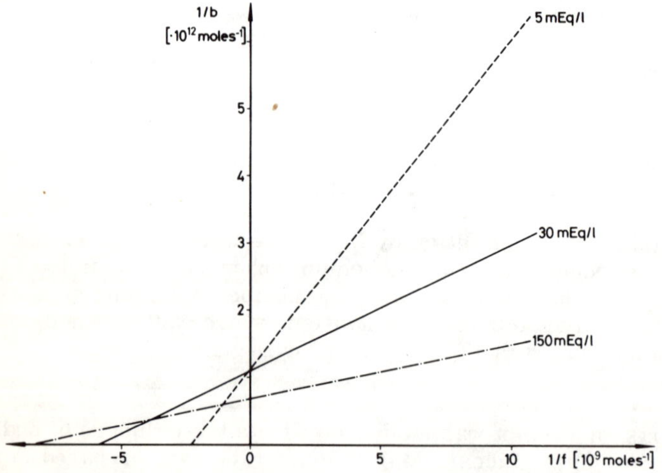



Fig. 3. The influence of sodium on the high affinity binding of phlorizin. Ordinate and abscissa as in Fig. 2.
Sodium concentrations of the medium :  150 mEq   -.-.-.-.
30 mEq   ————
5 mEq   ---------
The regression lines were calculated from all single data by the method of least squares

A considerable sodium chloride influx into the initially sodium-free perfusate increased the sodium concentration to an average of   29 mEq/l (range 1O to 5O mEq/l).  For technical reasons, only perfused distances between two collecting pipettes at least 300 micron downstream from the infusion pipettes were used for calculations.  After the proper correction for volume flux, the glucose reabsorption was essentially the same under both the high and low sodium condition.  The same results were obtained by using lithium chloride or raffinose instead of cholin chloride.

In order to achieve lower intraluminal sodium concentrations, the micro-perfusion method was used in combination with a simultaneous perfusion of the surrounding blood capillaries with artificial solutions.  Nevertheless, sodium, over the range of 30 to 150 mEq/l, had no influence on transtubular D-glucose transport.


## References


1. BAUMANN, K., HUANG, K. C.: Micropuncture and microperfusion study of L-glucose secretion in rat kidney. Pflügers Arch. 305 , 155-166 (1969).
2. BODE, F., BAUMANN, K., FRASCH, W., KINNE, R.: Die Bindung von Phlorrhizin an die Bürstensaumfraktion der Rattenniere. Pflügers Arch. 315, 53-65 (1970).
3. FRASCH, W., FROHNERT, P. P., BODE, F., BAUMANN, K., KINNE, R.: Competitive inhibition of phlorizin binding by D-glucose and the influence of sodium: a study on isolated brush border membrane of rat kidney. Pflügers Arch. 320, 265-284 (1970).
4. FROHNERT, P. P., HÖHMANN, B., ZWIEBEL, R., BAUMANN, K.: Free flow micropuncture studies of glucose transport in the rat nephron. Pflügers Arch. 315, 66-85 (197O).
5. KINNE, R., KINNE-SAFFRAN, E.: Isolierung und enzymatische Charakterisierung einer Bürstensaumfraktion der Rattenniere. Pflügers Arch. 3O8, 1-15 (1969).
6. LOESCHKE, K., BAUMANN, K.: Kinetische Studien der D-Glukoseresorption im proximalen Konvolut der Rattenniere. Pflügers Arch. 305 , 139-154 (1969).
7. LOESCHKE, K., BAUMANN, K., RENSCHLER, H., ULLRICH, K. J. mit einem mathematischen Anhang von FUCHS, G.: Differenzierung zwischen aktiver und passiver Komponente der D-Glukosetransports am proximalen Konvolut der Rattenniere. Pflügers Arch. 305, 118-138 (1969).
8. ULLRICH, K. J., FRÖMTER, E., BAUMANN, K.: Micropuncture and microanalysis in kidney physiology. In: Laboratory techniques in membrane biophysics (H. Passow and R. Stämpfli) Springer Verlag Berlin - Heidelberg - New York, p. 106-129 (1969).

# Views Dissenting with the "Gradient Hypothesis".
# Intestinal Sugar Absorption, Studies in vivo and in vitro

Harald Förster
Institut für vegetative Physiologie der Universität Frankfurt/Main, Germany

The model of cotransport or symport seems to be accepted for the absorption of glucose from the intestine (1, 2). According to this model (the "sodium gradient hypothesis") there exists a stoichiometric relation between the uptake of glucose and sodium from the intestinal lumen. The uphill transport of glucose is effected by the downhill transport of sodium (1, 2, 3). The sodium is transported out of the intestinal epithelium by the sodium pump whereas glucose leaves the intestinal epithelial cells at the basal side depending upon the concentration gradient. This model is based on conclusions derived mainly from in vitro investigations. Under in vivo conditions the outer surface of the intestine has to cope with different conditions, whereas the basal membrane is always surrounded by solutions of the same composition. On the other hand, in vitro the tissue is incubated in artificial buffer solutions and the blood supply is generally not intact. It is to be expected that there can be essential differences between the results of in vitro and of in vivo investigations.

If the symport model is correct, it is to be expected that inside the intestinal lumen a sodium concentration is maintained which is considerably higher than that in the epithelial cells. As the intracellular sodium concentration amounts to 40-60 meq/l (4, 5) one must assume that during active intestinal glucose absorption at least 100 meq/l of sodium are present in the intestinal lumen if the glucose concentration in the blood is twice the glucose concentration in the intestinal lumen. However, when rats are force-fed glucose solutions by means of a stomach tube, glucose is obviously absorbed though only 30-70 meq/l of sodium are measured inside the intestinal lumen.

In order to test the dependence of glucose absorption on sodium we have performed experiments with hyperglycemic animals (7, 8, 9). Under these conditions it is possible to measure uphill glucose absorption at luminal glucose concentrations in the range of the transport constant (half-saturation concentration) for glucose absorption. This was 20-30 mM for the system used, a value in good agreement with those published by other authors for in vivo experiments (see 8).

By means of an intravenous glucose infusion the blood glucose concentration was raised to 60-80 mM. When the intestinal lumen was perfused with glucose solutions of 27.8 mM, a net uptake of glucose was measured (8). This uphill transport of glucose is almost independent of the sodium concentration inside the intestinal lumen (7, 8). As long as isotonicity of the intestinal lumen is maintained, be it by NaCl, by magnesium sulfate, by cholin chloride or by polyols, such as xylitol, sorbitol or mannitol, to adjust the osmotic pressure of the intestinal content to the osmotic pressure of the blood, sugar uptake is the same (Fig. 1).

During the perfusion with sodium-free solutions, sodium is excreted into the intestinal lumen. The sodium concentration in the perfusate, i. e. after the passage

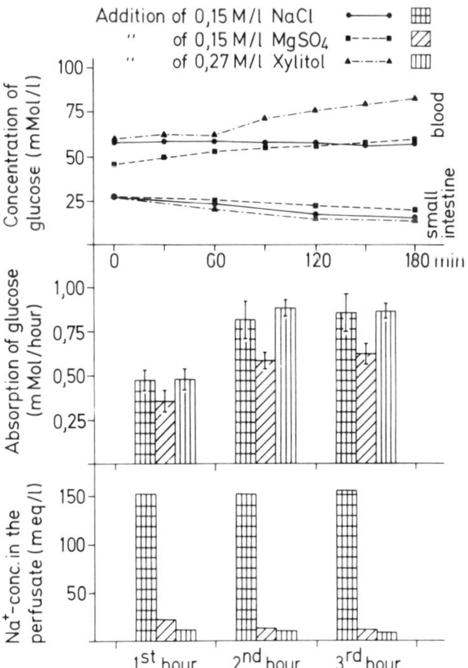

Fig. 1. The effect of sodium chloride and other substances on uphill glucose absorption from the small intestine of rats, in vivo. The concentration of blood glucose was elevated by means of an intravenous glucose infusion (taken from 8)

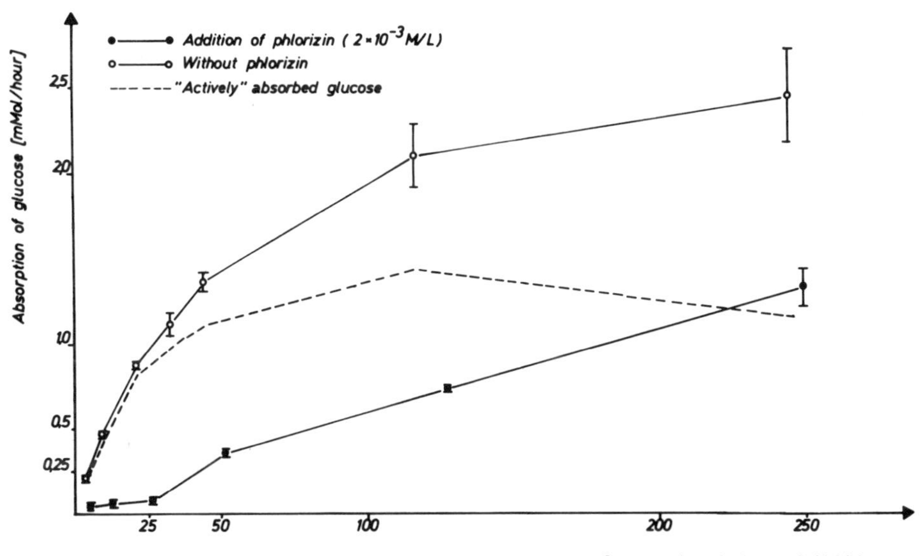



Fig. 2. Effect of phlorizin on the absorption rate of glucose, in vivo. The rate of disappearance of glucose from the perfused small intestine of rats was measured. Addition of 0. 15 M NaCl to the perfusion fluid (taken from 8)

of the intestine, amounts to 10-20 meq/l (7). This concentration is remarkably lower than the sodium concentration inside the epithelial cells. These results indicate that glucose is absorbed in vivo against its concentration gradient in the

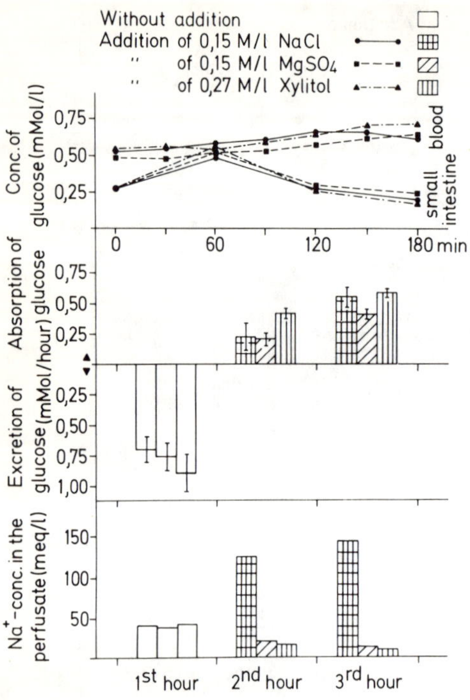



Fig. 3. Excretion of glucose into the small small intestine of rats during perfusion with hypo-osmolar solutions, in vivo. The concentration of blood glucose was elevated by means of an intravenous glucose infusion. During the last two hours osmotic-active substances were added to the perfusion fluid (taken from 8)

absence of an appropriate gradient of sodium ions. This sodium-independent glucose absorption is completely and reversibly inhibited by $2 \times 10^{-3}$ M phlorizin (Fig. 2).

In other experiments the intestinal lumen was perfused with solutions containing 27.8 mM of glucose without any other additions, i. e. with solutions hypoosmotic with respect to blood (Fig. 3). Under these conditions both glucose and sodium were excreted into the intestinal lumen. After passage of the intestinal lumen the sodium concentration of the perfusate had reached 40-50 meq/l, a concentration similar to the sodium concentration inside the intestinal epithelial cells. At the same time the glucose concentration in the perfusate had reached that in the blood. When in

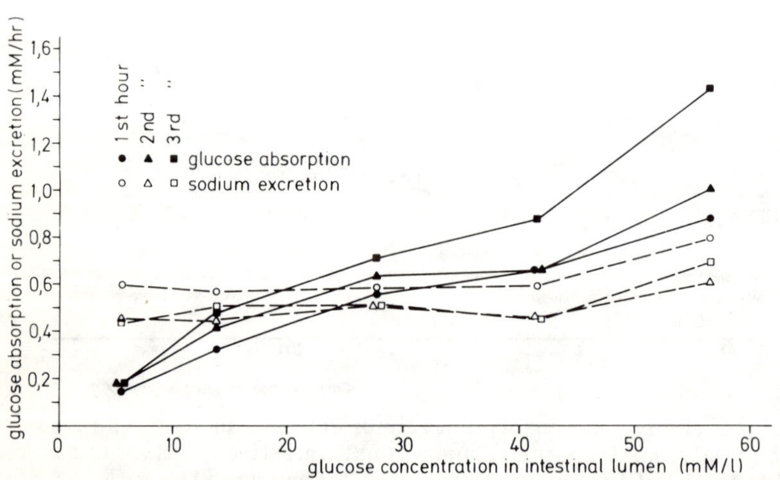



Fig. 4. Comparison between active intestinal glucose absorption and sodium excretion into the intestinal lumen in experiments performed with rats in vivo. The intestinal lumen was perfused with solutions containing glucose in varying concentrations (taken from 10)

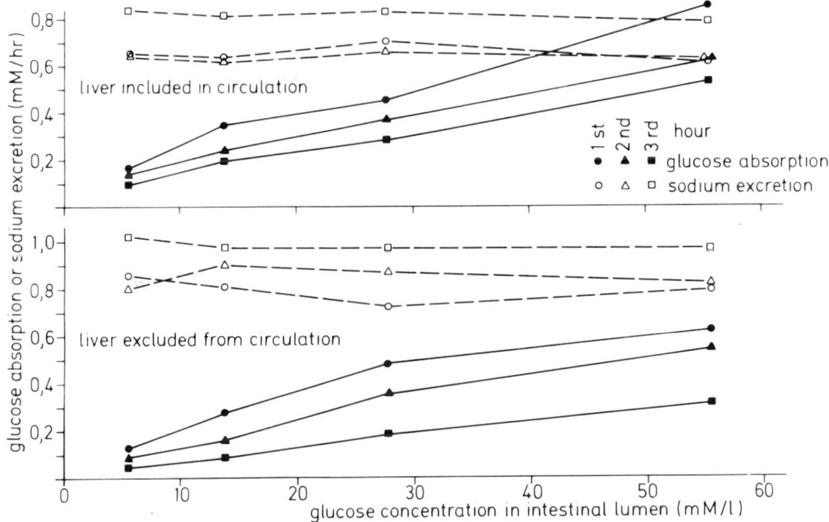



Fig. 5. Comparison between active intestinal glucose absorption and sodium excretion into the intestinal lumen in experiments performed in vitro with intact blood circulation. The intestinal lumen was perfused with solutions containing glucose in varying concentrations and 270 mM xylitol

the following periods the perfusion solution was made isoosmotic with the blood by NaCl, $MgSO_4$ or xylitol, glucose was transported uphill. For the active glucose transport in vivo the osmotic pressure inside the intestinal lumen seems to be more important than the sodium concentration.

We have further performed in vitro investigations using an intestinal preparation with intact blood circulation (9, 10). The results were almost the same as those of the in vivo experiments, except that in the former the transport capacity gradually decreased. In the living animal the glucose absorption is maximal; it is significantly decreased in experiments in vitro, even if the portal circulation of the liver is intact. The glucose absorption is further diminished when the liver is excluded from the portal circulation (Fig. 5). During our in vitro preparations the anoxia periods did not exceed 2 minutes. Under these conditions the glucose absorption remains almost independent of the sodium concentration inside the intestinal lumen. Moreover, the sodium excretion into the intestinal lumen is greater than in vivo. Also another index of the functional intactness of the preparation, the water absorption, decreases as compared to that of in vivo experiments. The ability to cope with changes of environment is obviously diminished in vitro, in spite of optimal conditions.

In vivo the transport constant for the glucose absorption is found to be 20-30 mM. Hence an increase of the glucose uptake is to be expected when the luminal glucose concentration is raised beyond this value. To study whether the absorption of glucose has an effect on the uptake of sodium from the intestinal lumen we have followed the excretion of $Na^+$ during glucose absorption. In these experiments glucose was actively transported as its concentration in the blood was always higher than that in the intestinal lumen. While the rate of the glucose absorption could be varied between 0 and 1.5 mmoles/h by varying the glucose concentration in the intestinal lumen, the excretion of sodium remained constant at 0.6 , eq/h (Fig. 4).

The same can be observed in in vitro experiments with maintained blood circulation: the absorption of glucose varies between 0.1 and 0.6 mM/h and the excretion of sodium remains constant at 0.8-0.9 meq/h (Fig. 5). These results exclude a correlation between the uptake of glucose and of sodium.

If for each molecule of glucose one sodium ion is transported (3), one would expect a decrease of sodium excretion by raising the rate of sugar uptake up to three times the maximal sodium excretion.

There are distinct differences between the transport constants measured under in vitro and in vivo conditions. The transport constant for glucose determined in vitro is only 1-2 mM, but in vivo, 20-100 mM. We have measured the arterioportal blood glucose concentration differences during the perfusion of the intestinal lumen with solutions containing glucose in different concentrations. Only if the glucose concentration exceeds 10 mM inside the intestinal lumen, glucose appears in the portal blood. A glucose arterioportal concentration difference of at least 50-100 mg/100 ml can be expected from oral glucose tolerance tests (11). This concentration difference is observed only when the intestinal lumen is perfused with solutions containing 100-300 mM glucose (Fig. 6). These results argue against the physiological importance of a transport process with a transport constant as low as 1 mM.

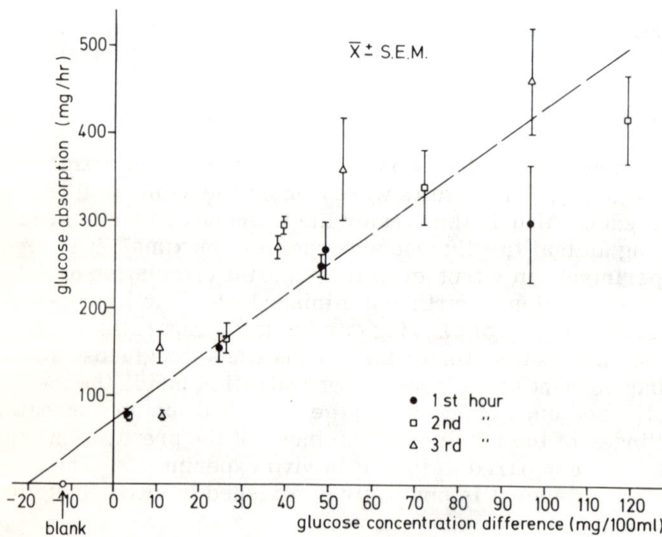


Fig. 6. Relation between intestinal glucose absorption and arterio-portal glucose concentration differences in studies performed with rats. Glucose absorption was modified by varying the glucose concentration in the intestinal lumen

## Conclusions

Our in vivo and in vitro experiments indicate the existence of an intestinal transport process independent of sodium inside the intestinal lumen (see also 12, 13). This transport process has a $K_T$ of 20-30 mM and is obviously responsible for the increase in glucose concentration in the portal blood during the intestinal glucose absorption. In addition there may exist also a sodium-dependent system for glucose absorption which is demonstrated only under special in vitro conditions. However, because of its low $K_T$ this transport system is unlikely to be of physiological importance for the intestinal absorption of glucose.

## Acknowledgements

This study was supported by the Deutsche Forschungsgeminschaft.

## References


1. CRANE, R. K.: Fed. Proc. 24, 1000 (1965).
2. SCHULTZ, S. G., CURRAN, P. F.: Physiol. Rev. 50, 637 (1970).
3. CRANE, R. K.: Biochem. Biophys. Res. Commun. 17, 481 (1965).
4. BOSACKOVA, J., CRANE, R. K.: Biochim. Biophys. Acta 102, 436 (1965).
5. SCHULTZ, S. G., FUIZ, R. E., CURRAN, P. F.: J. Gen. Physiol. 53, 362 (1966).
6. CSAKY, T. Z., HO, P. M.: J. Gen. Physiol. 50, 113 (1966).
7. FÖRSTER, H., KAISER, W., MEHNERT, H.: Klin. Wschr. 43, 844 (1965).
8. FÖRSTER, H., MENZEL, B.: Z. Ernährungswiss (in press).
9. FÖRSTER, H., HOOS, I.: First Europ. Biophysics Congr. E VIII/30 (1971).
10. FÖRSTER, H., HOOS, I.: Hoppe-Seylers Z. Physiol. Chem. 351, 1302 (1970).
11. FÖRSTER, H., HASLBECK, M., MEHNERT, H.: Diabetes (in press).
12. CSAKY, T. Z.: Fed. Proc. 22, 3 (1963).
13. KIMMICH, G. A.: Biochemistry 9, 3669 (1970).

# Two Modes of Sodium Extrusion from Dog Kidney Cortex Slices

J. W. L. Robinson
Département de Chirurgie Expérimentale, Hôpital Cantonal Universitaire,
Lausanne, Suisse

## Abstract

The uptake of glycine and ß-methyl-glucosides by dog kidney cortex slices against
a concentration gradient is entirely sodium-dependent. However, this active transport cannot be totally abolished by ouabain, even at high concentrations. Besides
being partially inhibited by ouabain, it is also partially inhibited by ethacrynic acid.
The effects of the two drugs are additive, and accumulative uptake is abolished
in the presence of high concentrations of both. These results are consistent with
the hypothesis of two separate sodium pumps in the dog kidney cortex cell, each
being inhibited by one of the drugs.

## Introduction

One of the cornerstones of the hypothesis linking the active transport of sugars
and amino acids to the sodium gradient across the cell membrane has been the effect of classical sodium-pump inhibitors on these active transport mechanisms.
It was argued (Crane, 1962) that since cardiac glycosides inhibit both sodium transport out of the cell and sugar or amino acid transport into it, there must be some
vital link between the two processes. It was thus proposed that the maintenance of
the sodium gradient across the luminal membrane of the intestinal cell was obligatory for the accumulative uptake of sugars or amino acids into the cell; on inhibiting the sodium pump responsible for sustaining this gradient, there was a simultaneous inhibition of sugar or amino acid transport.

This argument, which has been upheld by experimental data of many types (Schultz
and Curran, 1970), can also be used for the study of the sodium-pump mechanisms
of the intestinal or renal cell. In order that all active uptake of nonelectrolytes may
be abolished by destroying the sodium gradient across the cell membrane, all sodium-pumping mechanisms must be fully inhibited. If only one pump is inhibited,
and other(s) continue to extrude sodium from the cell, only partial inhibition, at
best, of the non-electrolyte uptake can occur. In an attempt to obtain further evidence on sodium-pumping mechanisms of the proximal tubular cell of the dog kidney cortex, we have undertaken studies on the inhibition of sugar or amino acid
uptake in tissue slices by different inhibitors of sodium-pumping mechanisms. The
results provide confirmatory evidence in favour of the dual pump hypothesis (Whittembury and Proverbio, 1970), according to which one pump, a $Na^+$-$K^+$-ATPase
sensitive to ouabain, controls the ionic content of the cell, whereas a second pump,
insensitive to ouabain but inhibited by ethacrynic acid, regulates the movement
of sodium and water across the cell.

## Methods

Slices of 0.5 mm thickness were prepared from the renal cortex of healthy mongrel dogs by means of a home-made, spring-loaded guillotine. They were incubated for one hour in a $^{14}$C-labelled solution of glycine or ß-methyl-glucoside in Krebs bicarbonate buffer, after which the individual slices were rinsed, weighed and dissolved in o.1 ml 30% KOH for counting in a liquid scintillation counter in a toluene/ethanol medium (Robinson and Felber, 1965). Aliquots of the incubation medium were counted under identical conditions. Ouabain and/or ethacrynic acid were added to the incubation medium at concentrations dictated by the experiment in question. For experiments on sodium-dependence of the uptake of the substrates, all sodium ions of the buffer were replaced by equimolar quantities of potassium or choline ions.

## Results

The substrates, glycine and ß-methyl-glucoside, were chosen because the sodium-dependency of their transport in kidney cortex slices or other species had already

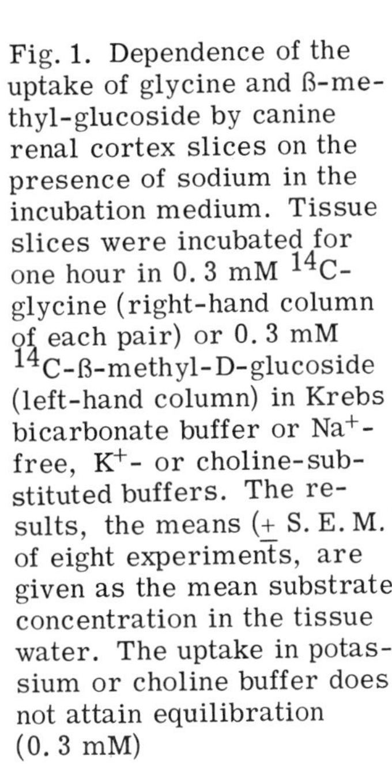

Fig. 1. Dependence of the uptake of glycine and ß-methyl-glucoside by canine renal cortex slices on the presence of sodium in the incubation medium. Tissue slices were incubated for one hour in 0.3 mM $^{14}$C-glycine (right-hand column of each pair) or 0.3 mM $^{14}$C-ß-methyl-D-glucoside (left-hand column) in Krebs bicarbonate buffer or Na$^+$-free, K$^+$- or choline-substituted buffers. The results, the means ($\pm$ S.E.M.) of eight experiments, are given as the mean substrate concentration in the tissue water. The uptake in potassium or choline buffer does not attain equilibration (0.3 mM)

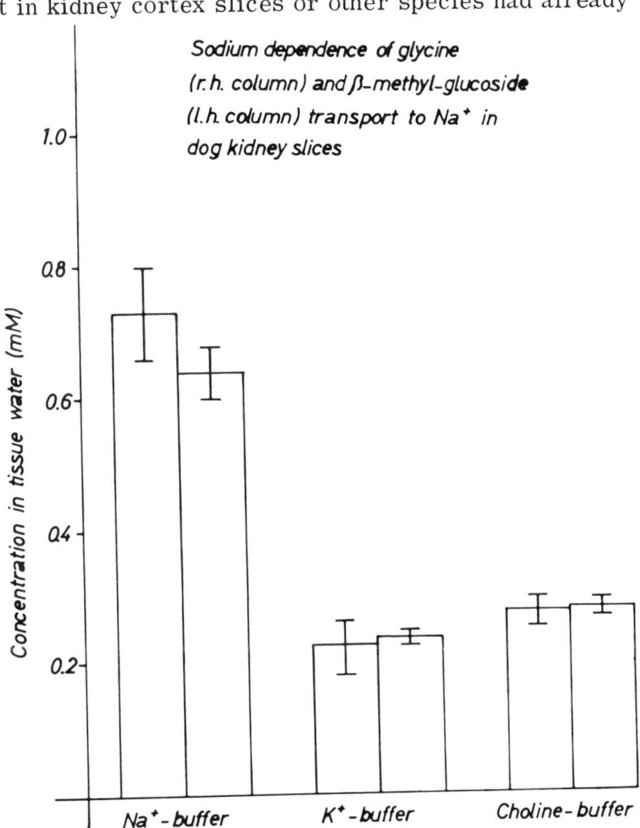


been demonstrated (Fox et al., 1964; Kleinzeller, 1970). The results in Fig. 1 confirm that in dog renal cortex slices, no accumulative uptake occurs when sodium is replaced by potassium or choline. Chromatography demonstrated that these substrates were not metabolised by this tissue.

Although the uptake of glycine by this tissue is sensitive to ouabain, the entire active component of uptake cannot be eliminated, even at very high concentrations of the glycoside. Fig. 2 shows the dose-response curve of its inhibition by ouabain;

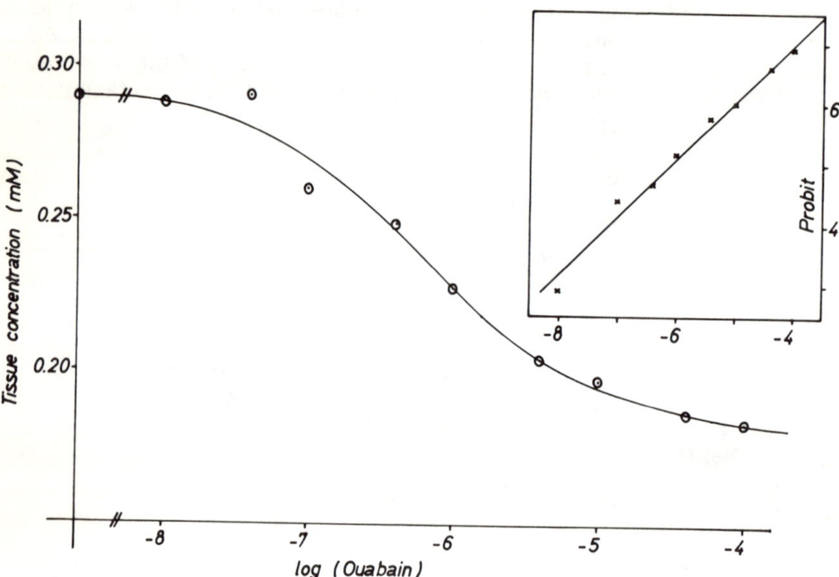



Fig. 2. Sensitivity to ouabain of glycine transport by dog kidney cortex slices. Results are the means of four experiments with different dogs. The ordinate represents the mean substrate concentration in tissue water after one hour's incubation in 0.1 mM $^{14}$C-glycine. The results were subjected to probit analysis and the best fit was found by iteration, using different values for the asymptote and testing the linearity of the resulting probit lines. The best probit line (based on an asymptote of 0.181 mM) is shown in the inset, the ordinate representing the probit of the proportion of the maximal effect, and the abscissa (as in the main graph) being the logarithm of the molar concentration of ouabain

probit analysis has indicated that the most likely asymptote for the curve obtained is 0.181 mM, that is to say that 57% of the active uptake can be inhibited by ouabain. The semi-maximal inhibition of the ouabain-sensitive component of glycine uptake occurs at an inhibitor concentration of 0.63 $\mu$M, which is in relatively close agreement with the sensitivity of the Na$^+$-K$^+$-ATPase of this tissue to ouabain - 2 $\mu$M according to Nelson and Nechay (1970).

The uptake of glycine or ß-methyl-glucoside is also sensitive to ethacrynic acid. In Fig. 3, the effect of this drug at high concentrations (2 mM) is presented, and it can be seen that it also inhibits rather more than 50% of the active uptake of ß-methyl-glucoside. However, when both ouabain and ethacrynic acid are added together to the incubation medium at sufficient concentrations, no significant active uptake of the substrate remains. Note that addition of cysteine tends to prevent the inhibitory action of ethacrynic acid, so the drug is presumably acting in this case by means of its affinity for -SH groups.

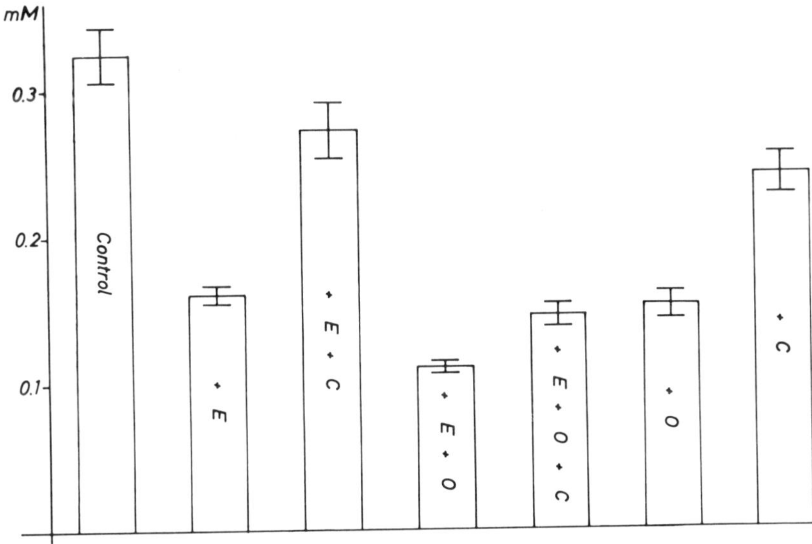



Fig. 3. Effect of ethacrynic acid and ouabain on the transport of ß-methyl-gluco-side. Antagonistic effect of L-cysteine. Slices were incubated for one hour in 0.1 mM $^{14}$C-ß-methyl-glucoside in the presence of the various inhibitors or mix-tures (E = 2 mM ethacrynic acid; 0 = 0.5 mM ouabain; C = 2 mM cysteine). Re-sults presented as the means of six experiments ($\pm$ S.E.M.) and expressed as mean substrate concentration in tissue water. The difference between the effect of ouabain alone and of ouabain + ethacrynic acid is significant at the 0.5% level

The additivity of the effects of the two drugs is confirmed by the curve shown in Fig. 4. This is a dose-response curve of the inhibition of glycine uptake by ouabain, analogous to that shown in Fig. 2, but in the presence of a constant high concentration of ethacrynic acid. In this case, the asymptote of the curve is close to a distribu-tion ratio of unity. A 50% inhibition is obtained at a ouabain concentration of 0.8 $\mu$M, which agrees closely with that obtained in the absence of ethacrynic acid (Fig. 2).

## Discussion

"The present work was undertaken on the assumption that the transport capacity for glycine or ß-methyl-glucoside provides a satisfactory measure of the situation regarding the sodium gradient in the presence of inhibitors of the sodium pumps. In view of findings with epithelial tissues reviewed elsewhere (Schultz and Curran, 1970), and of the complete sodium-dependency of the transport of the substrates chosen (Fig. 1), this assumption appears justified".

The results presented here are consistent with the hypothesis that the proximal tubular cell possesses two distinct sodium pump mechanisms, as was proposed

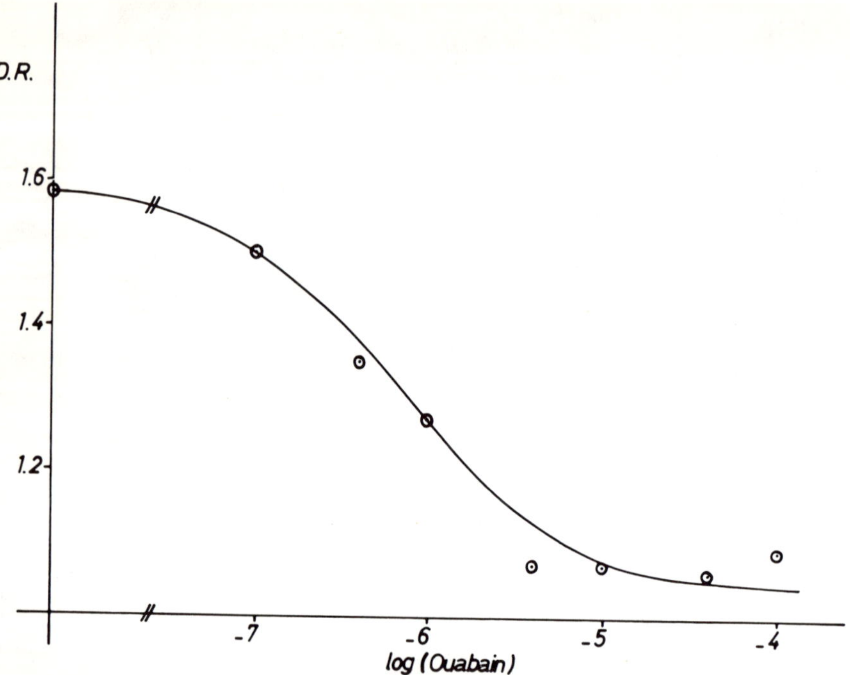

Fig. 4. Inhibition by ouabain of the uptake of glycine by dog renal cortex slices in the presence of ethacrynic acid. Slices were incubated for 1 hour in 0. 1 mM $^{14}$C-glycine + 1 mM ethacrynic acid in the presence of different concentrations of ouabain. The results are the means of four experiments, and are expressed as the ratio between intra- and extra-cellular concentrations ("distribution ratio"), calculated on the basis of 83% tissue water, and 18. 3% extracellular space

as a result of investigations of ion movements in guinea pig kidney cortex slices (Whittembury and Proverbio, 1970). One pump is sensitive to ouabain, is non-electrogenic (Proverbio and Whittembury, 1969), and is probably mediated by a $Na^+$-$K^+$-ATPase. Its main function appears to lie in the regulation of cellular ion content, since it is primarily concerned with a stoichiometric exchange of intra-cellular sodium ions for extracellular potassium. The second pump, on the other hand, is independent of potassium, electrogenic and responsible for movement of sodium in conjunction with chloride. Since it is coupled to water movement, it probably mediated bulk flow across the cells. It is refractory to ouabain, but inhibited by ethacrynic acid; it appears to be independent of the action of a $Na^+$-$K^+$-ATPase (Proverbio et al., 1970).

Studies on perfused toad kidneys in vivo (Whittembury and Fishman, 1969), using perfusates of different ionic compositions or containing the inhibitors, ouabain and/or ethacrynic acid, have provided results which cannot easily be reconciled with a single $Na^+$-$K^+$-ATPase pump in the basal membrane of the proximal tubular cell.

These findings of Whittembury and his associates have been upheld by various re-sults from other laboratories: Kessler et al. (1968) observed that there was no correlation in the living dog between ATP turnover and sodium transport in the kidney; Nelson and Nechay (1970) infused ouabain into the renal artery, totally inhibited the $Na^+$-$K^+$-ATPase of the renal cortex, but failed to abolish sodium transport completely; and Munday et al. (1971) studied the effect of angiotensin on sodium transport in rat kidney cortex slices, and observed a stimulation of sodium transport, even in the presence of high concentrations of ouabain, which was accompanied by an inhibition of potassium movement. Thus there appears to exist a second sodium pumping mechanism in the kidney, unrelated to the $Na^+$-$K^+$-ATPase. The enzymatic mechanism involved in the functioning of this pump is completely unknown and its elucidation presents an important challenge to membrane physiologists.


## References


CRANE, R. K.: Hypothesis for mechanism of intestinal active transport of sugars. Fed. Proc. 21, 891-895 (1962).

FOX, M., THIER, S., ROSENBERG, L., SEGAL, S.: Ionic requirements for amino acid transport in the rat kidney cortex slice. I. Influence of extracellular cations. Biochim. biophys. Acta 79, 167-176 (1964).

KESSLER, R. H., LANDWEHR, D., QUINTANILLA, A., WESELEY, S. A., KAUFMANN, W., ARCILA, H., URBAITIS, B. K.: Effects of certain inhibitors on renal sodium reabsorption and ATP specific activity. Nephron 5, 474-488 (1968).

KLEINZELLER, A.: The specificity of the active sugar transport in kidney cortex cells. Biochim. biophys. Acta 211, 264-276 (1970).

MUNDAY, K. A., PARSONS, B. J., POAT, J. A.: The effect of angiotensin on cation transport by rat kidney cortex slices. J. Physiol. (London) 215, 269-282 (1971).

NELSON, J. A., NECHAY, B. R.: Effects of cardiac glycosides on renal adenosine triphosphatase activity and $Na^+$ reabsorption in dogs. J. Pharmacol. exp. Therap. 175, 727-740 (1970).

PROVERBIO, F., ROBINSON, J. W. L., WHITTEMBURY, G.: Sensitivities of $(Na^++K^+)$-ATPase and $Na^+$ extrusion mechanism to ouabain and ethacrynic acid in the cortex of the guinea-pig kidney. Biochim. biophys. Acta 211, 327-336 (1970).

PROVERBIO, F., WHITTEMBURY, G.: Potencial de la membrana y bombas de Na en celulas de la corteza renal de cobayo. Acta cient. venez. 20, 85 (1969).

ROBINSON, J. W. L., FELBER, J. -P.: Compartments of the uptake of amino-acids by intestinal fragments during in vitro incubation. Gastroenterologia (Basel) 104, 335-342 (1965).

SCHULTZ, S. G., CURRAN, P. F.: Coupled transport of sodium and organic solutes. Physiol. Revs. 50, 637-718 (1970).

WHITTEMBURY, G., FISHMAN, J.: Relation between cell Na extrusion and transtubular absorption in the perfused toad kidney: the effect of K, ouabain and ethacrynic acid. Pflügers Arch. -Europ. J. Physiol 307, 138-153 (1969).

WHITTEMBURY, G., PROVERBIO, F.: Two modes of Na extrusion in cells from guinea-pig kidney cortex slices. Pflügers Arch. -Europ. J. Physiol. 316, 1-25 (1970).

# General Comment

K. P. Wheeler

Department of Biological Chemistry, University of Michigan, Ann Arbor, Michigan 48104, USA

If we temporarily ignore the details of models, stoichiometry, kinetics, thermodynamics and interaction among sugars and amino acids, we are left with the simple fact that the movements of a variety of organic solutes are somehow associated with the distribution and movement of $Na^+$, and perhaps $K^+$, through a number of biological membranes. The simplest conclusion, surely, is that there is a common molecular basis for these observations. Now the direction of the associated cation flux is generally thought to be "downhill", and other evidence excludes the direct involvement of the ubiquitous membrane ATPase system in these linked fluxes, so that we could, perhaps, reverse our usual approach to the problem by saying that the organic solutes somehow affect what is commonly called the "passive" permeability properties of cell membranes. Evidence is accumulating, at least with the red blood cell, that shows that the "passive" permeability properties of the cell membrane are far from being intrinsic and static, but are intimately connected with and controlled by numerous factors, in particular $Ca^{++}$ concentration, metabolism, and the oxidation state of membrane -SH groups (see Whittam, R. and Wheeler, K. P., Ann. Rev. Physiol. <u>32</u>, 21 (1970)). Appropriate manipulation of these factors can produce marked changes in membrane permeability to $Na^+$ and $K^+$, often preferentially to one of these, so that it might be profitable to look for some common molecular basis along these lines.

# Sodium Activation of Intestinal Sugar and Amino Acid Transport: A General or an Individual Effect?

Francisco Alvarado
Department of Physiology, School of Medicine, University of Puerto Rico,
San Juan, Puerto Rico

In the course of a more comprehensive study of the kinetics of sugar transport in guinea-pig small intestine, it was found that changing the ratio of the monovalent cations, sodium and potassium, had qualitatively different kinetic effects on the transport of the two isomers, $\alpha$ - and ß-methylglucopyranoside. Such result seems to shed a new light on the interactions between sugar and amino acid transport systems in the small intestine, and particularly on the problem of whether the sodium-binding sites involved in the activation of sugar transport are the same as those involved in amino acid transport activation.

## Methods

Transport was determined by the "tissue accumulation method" as described previously (1-3). Briefly, 5 mm section from the everted jejunum and proximal ileum of unfasted guinea pigs (200 to 500 grams) were incubated in buffers of the composition indicated below, in the presence of $^{14}$C-labeled substrate (Calbiochem), diluted for the appropriate initial concentrations with cold substrate (Sigma); mannitol was added when necessary to maintain isoosmolarity. Radioactivity accumulated within the tissues was determined after digestion with KOH (3). For more details, see legend of Figure 1.

Buffer solutions. In all these experiments, a phosphate buffer was used with a constant total concentration of Na$^+$ plus K$^+$ (151. 6 m-equivalents per liter), but changing the ratio of Na$^+$ to K$^+$ between two extremes. The first buffer, of composition similar to that of Krebs and Henseleit (4), had a Na/K ratio of 25. 54; the second buffer, otherwise identical, had a Na/K ratio of 0. 47.

Calculation of results. As previously described (1-3), results are expressed as velocities (v), i. e., as  moles of substrate accumulated per ml tissue water per minute of incubation at 37$^{\text{O}}$. Water content was determined to be about 80% of the total net weight, and did not change during the short incubations used (up to 10 min). The results were always corrected for the extracellular space, measured with $^{3}$H-mannitol in separate pieces of tissue incubated in parallel. This space was quite constant and amounted to about 7. 5% of the external substrate concentration, in 5 minutes. As a check of this value, the kinetic analysis of Inui and Christensen (5) was applied to the experimental data to verify the homogeneity of the system and to calculate the extracellular space, as described previously (6). Both methods of calculation of the extracellular space gave essentially identical results.

Calculation of the kinetic constants. The reciprocals of the velocity data were plotted according to Lineweaver and Burk (7) against the reciprocals of the average substrate concentrations, i. e., the mean of (S) before and after incubation. The pooled data from several experiments were averaged and fed to a Hewlett-Packard

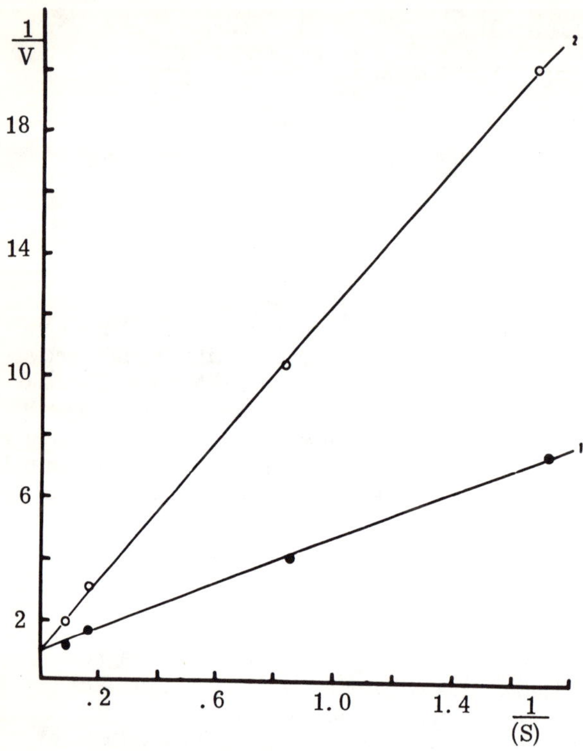

Fig. 1. Effect of the Na$^+$/K$^+$ ratio on the transport of methyl-$\alpha$-glucoside by guinea pig small intestine. Pieces of everted intestine were incubated at 37$^\circ$ in 4. 8 ml oxygenated phosphate buffer plus 0. 2 ml of a mixture of cold $\alpha$-glucoside and mannitol to give the appropriate initial substrate concentrations (0. 6 to 12. 0 mM): All flasks also had an identical amount of tracer $\alpha$-glucoside-$^{14}$C. Each flask received four pieces of randomized tissue (about 200 mg total net weight). After the incubations, these pieces were separated in groups of 2 for KOH degestion and separate radioactivity counting. The experimental points shown, plotted according to Lineweaver and Burk (7), are averages of 4 separate experiments, each experiment including 4 separate determinations per point. In three experiments, the incubations were for 5 minutes and in the fourth, for 10 minutes. The results, expressed as $\mu$ moles transported per minute, did not differ significantly and were therefore pooled. The Na$^+$/K$^+$ ratios in the buffers were: curve 1, 25. 54 and curve 2, 0. 47

9100B computer to calculate by the least-square method the straight lines shown in the figures. All the lines shown in the figures gave correlation coefficients greater than 0. 999. The constants, $V_{max}$ and $K_m$, were calculated from the computer figures for the intercept and slope, respectively, of the same lines.

## Results

Figure 1 is a Lineweaver-Burk (7) plot of the effect of a changing Na/K ratio on the kinetics of $\alpha$-methylglucoside transport. It is clearly seen that two straight lines are obtained, as expected for a system conforming to Michaelis-Menten kinetics. Since these two lines have a common intercept, but with a higher slop at the lowest Na/K ratio used, it seems clear that, as this ratio increases, there is an increase in the apparent affinity of the system for the $\alpha$-glucoside, without any significant effect on the rate of translocation: we may call this a K effect[+]. Similar K effects have been described for the sugar transport system in hamster small intestine; and for amino acid transport in the rabbit (for review, see ref. 8).

[+] Although it might be premature to try to adapt to this system the allosteric model of Monod et al. (9), we may be justified in using their terminology.

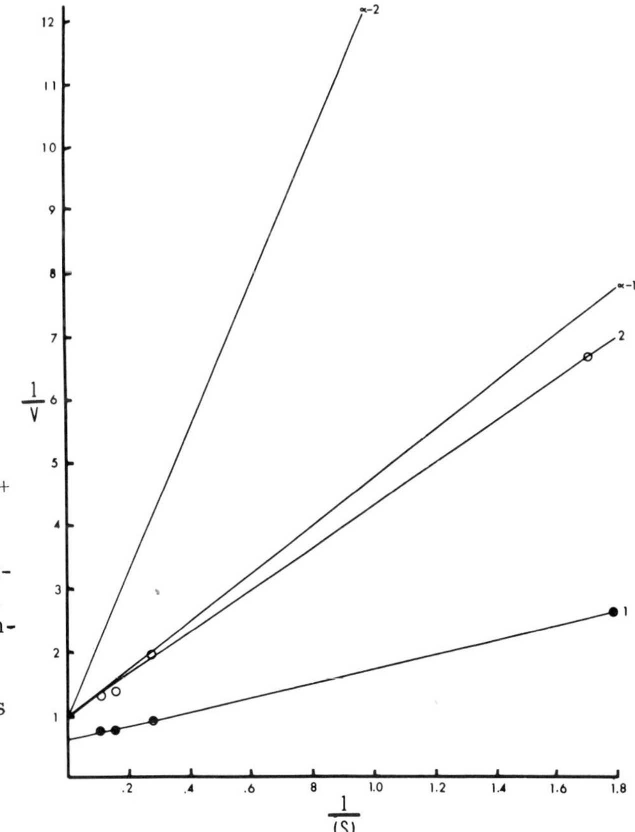



Fig. 2. Effect on the $Na^+/K^+$ ratio on the transport of methyl-ß-glucoside by guinea pig small intestine. Experimental conditions as in Fig. 1. Results are the summary of 3 separate experiments, all with incubation times of 5 min. The curves of the previous experiment (curves α -1 and α -2) are included for direct comparison

Figure 2 shows a similar experiment using ß-glucoside as the substrate; the lines obtained for the α -isomer in the previous experiment are also included to permit a direct visual comparison between the two results. Again straight lines are obtained, but the overall result is qualitatively quite different, the effect being mixed . On the one hand, and similar to the case of the α -isomer, an increase in the Na/K ratio leads to an increase in the apparent affinity of the transport system for the sugar (K effect). On the other hand, however, there is an additional effect (a V effect) not observed with the α -isomer, i. e., an increase in the rate of translocation indicated by a significant decrease in the intercept.

Thus, the effect of increasing the Na/K ratio in the mucosal solution has, in the case of the ß-glucoside, features of the effect quoted before for the systems of sugar transport in hamster and of amino acid transport in rabbit small intestine, plus features of the effect described for sugar transport in rabbit intestine (8); i. e., an increase in the apparent affinity (decrease in $K_m$) in one case, and an increase in $V_{max}$ in the other.

These results could perhaps be interpreted as suggesting that the ß-isomer, but not the α -isomer, is transported by two separate mechanisms: one common for both isomers, at which level a straight competition for a single binding site between both substrates would occur; and a second system specific for the ß-isomer, that would be unoperative at the lowest Na/K ratio used, but that would become active

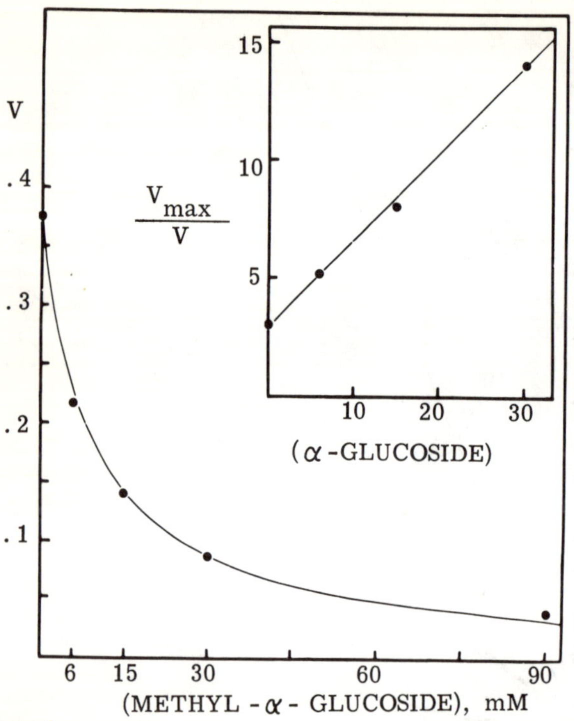



Fig. 3. Effect of increasing $\alpha$-glucoside concentrations on the transport of ß-glucoside. Incubations were for 5 min in 3. 5 ml phosphate buffer (Na/K = 25. 54) supplemented to a total volume of 5 ml with ß-glucoside-$^{14}$C to a constant initial concentration (0. 9 mM); $\alpha$-glucoside to give the initial concentrations shown; and mannitol to maintain isoosmolarity. In the insert, the results are plotted according to Thorn (11). From the slope of this line, calculated by least squares, a $K_i$ value of 6. 1 mM for the $\alpha$-glucoside was calculated, in reasonable agreement with the $K_m$ (4. 2 mM) calculated for this sugar in the Table

at the highest Na/K ratio, thus accounting for the increase in the $V_{max}$ of only the ß-isomer.

To check this possibility, competition studies were carried out. One of several experiments, all of which gave similar results, is shown in Figure 3. The results, calculated according to the methods of Inui and Christensen (5), Dixon (10) and Thorn (11) (only the last one is represented in the inset of figure 3), clearly indicate that, both at low and at high Na/K ratio, the two methylglucosides share a single binding site for which they compete according to regular kinetics of fully competitive inhibition.

These results, therefore, demonstrate that allosteric activation by $Na^+$ at the level of a single carrier may consist, according to the substrate bound to the allosteric substrate site, in either an effect on affinity (K effect) or in a mixed effect on both affinity and transport rate (K plus V effect). Theoretically, an effect exclusively on $V_{max}$ is also conceivable, and having found the above two effects (on affinity and mixed), using two substrates so closely related as the two methylglucosides, it seems logical to predict that, if a more thorough exploration is undertaken using other sugars among the list of substrates for the sugar transport system, an effect on $V_{max}$ only might also be found.

## Discussion

Sodium ions are an absolute requirement for the uphill transport, in the small intestine, of a variety of substances that includes sugars, amino acids, pyrimidines, bile salts, sulfate ions, and ascorbic acid (8).

This unusual requirement for the same ion among such a diverse series of transport systems, calls for the existence of some common link. If we accept the evidence thus far accumulated for the case of sugar and amino acid transport systems, the best-studied among the above list, then the common link must occur at the membrane level where the sodium ions are believed to interact with the specific transport proteins or carriers (13).

Intestinal membrane carriers are believed to be at least bifunctional; i.e., they posses a substrate-binding site plus a sodium-binding site. These are allosterically interrelated in such a manner that $Na^+$ stimulates transport of both sugars and amino acids, and $Na^+$ transport is in turn stimulated by the presence of either sugars or amino acids, these interactions taking place from the external side of the membrane, the "mucosal fluid".

The problem, however, still remains of whether the sugar (S) and the amino acid (A) sites are independent or interrelated (12). Two main hypotheses have been put forward to explain the situation, as schematized in Figure 4. In one hypothesis (Fig. 4-I), S and A sites are thought to be entirely separate, each one being independently associated to one sodium site (it is not implied that the ratios of sites are 1 to 1 in all cases; for more details see ref. 6). According to such hypothesis, S and A transport systems are "different" or "separate", although they are "similar" or "parallel" since both have in common a requirement for $Na^+$. This ion, then, would act in an "individual" rather than in a "general" manner.

In another hypothesis (Fig. 4, II and III), the same sodium-binding site associates with one S and with one A site, i.e., $Na^+$ is a "general" rather than an "individual" activator. Such a model is one of a "polyfunctional carrier" in which direct interactions between the S and A sites may either occur (Fig. 4-III) or be absent (Fig. 4-II).

All the evidence thus far available from studies on hamster intestine appears to support the last hypothesis ($Na^+$ as a general activator; see ref. 6), but there is much disagreement with regard to the applicability of this hypothesis to other spe-

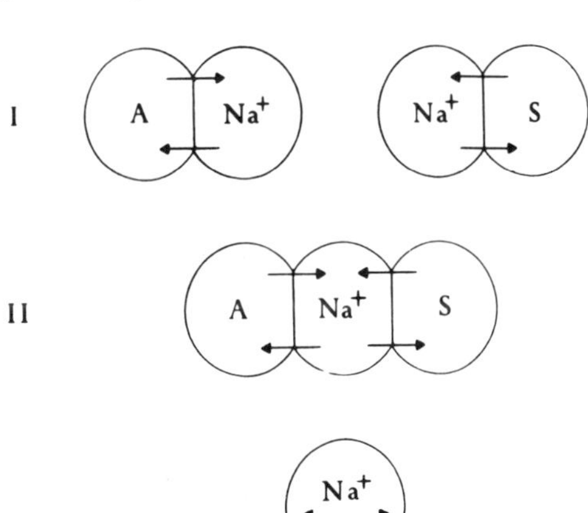


Fig. 4. Schematic representation of the possible ways of interaction between membrane sites for sugars, amino acids and $Na^+$, in the assumption that $Na^+$ is an individual (I) or a general (II, III) activator of sugar and amino acid transport. Taken from ref. 6, with permission of the publisher

cies (8, 12). In their thorough study of the kinetics of sugar and amino acid transport in rabbit ileum, Schultz, Curran and their colleages (8) found that $Na^+$ had qualitatively different kinetic effects on either system: a K effect in the case of amino acid transport and a V effect in the case of sugar transport. Such results led these workers to postulate quite different mechanisms for the sodium-dependent transport of these two families of compounds. Specifically, Curran has addressed himself to the question of whether $Na^+$ is an individual or a general activator with the affirmation that their work has led them "to postulate such widely different roles for sodium in the two transport systems that it is very difficult to see how one sodium site could be common to both"(14).

The observations summarized in this paper, however, appear to make this kind of reasoning untenable. In a single carrier system, that for sugar transport in the guinea-pig small intestine, $Na^+$ appears to have a clear-cut K effect when a certain substrate, $\alpha$-glucoside, is used. But the $Na^+$ effect becomes mixed, i. e., involves both a K and a V effect, when the substrate is the ß-glucoside. Since both sugars compete for the same binding site, it seems evident (and is in keeping with what one would have expected, based on theoretical grounds, see ref. 9) that any theoretical model aiming to explain sodium activation of a transport system should leave room for consideration of both K and V effects, these depending to a large extent on the nature of the substrate analog used to test the system.

Therefore, it may be concluded that all the results thus far available in either hamster, rabbit or other mammals, are quite compatible with the interpretation that activation of sites S and A can involve a single (common) sodium-binding site, i. e., $Na^+$ can be a general rather than an individual activator of sugar and amino acid transport systems in small intestine. The implications of this concept in regard to the structure and function of biological membranes seem obvious.


## Acknowledgements

Aided by grant AM-13118 and Research Career Development Award 5-KO-AM 42383 from the National Institute of Arthritis and Metabolic Diseases, U. S. P. H. S. The skilful technical assistance of Mrs. Beatriz Robillard is greatly appreciated.


## References


1. CRANE, R. K. , MANDELSTAM, P. : Biochim. Biophys. Acta 45, 460 (1960).
2. ALVARADO, F. : Biochim. Biophys. Acta 109, 478 (1965).
3. ROBINSON, J. W. L. , ALVARADO, F. : Pflügers Arch. 326, 48 (1971).
4. KREBS, H. A. , HENSELEIT, K. : Hoppe-Seylers T. Physiol. Chem. 210, 33 (1932).
5. INUI, Y. , CHRISTENSEN, H. N. : J. Gen. Physiol. 50, 203 (1966).
6. ALVAROADO, F. : in: Intestinal transport of electrolytes, amino acids and sugars (eds. W. McD. Armstrong and A. S. Nunn), Ch. C. Thomas, Springfield, Ill. 1971, pp. 281-318.
7. LINEWEAVER, H. , BURK, D. : J. Amer. Chem. Soc. 56, 658 (1934).
8. SCHULTZ, S. G. , CURRAN, P. F. : Physiol. Rev. 50, 637 (1970).
9. MONOD, J. , WYMAN, J. , CHANGEUX, J. -P. : J. Mol. Biol. 12, 88 (1965).
10. DIXON, M. , WEBB, E. C. : Enzymes, 2nd ed. New York, Academic, 1964.
11. THORN, M. B. : Biochem. J. 54, 540 (1953).
12. ALVARADO, F. : J. Clin. Invest. 23, 824 (1970).

13. ALVARADO, F.: Bol. R. Soc. Espanola Hist.Nat. (Biol.) **68**, 33 (1970).
14. CURRAN, P. F.: in Discussion of ref. 6, p. 317.

Table

Kinetic constants for the transport of methylglucopyranosides by slices of guinea-pig small intestine. Summary of data from Figures 1 and 2. A detailed statistical analysis will be published elsewhere.

| $Na^+/K^+$ ratio | 0.47 | | 25.6 | |
| --- | --- | --- | --- | --- |
| | $V_{max}$ | $K_m$ | $V_{max}$ | $K_m$ |
| $\alpha$-glucoside | $1.0 \pm 0.10$ | 10.6 | $1.0 \pm 0.1$ | 4.2 |
| ß-glucoside | $1.0 \pm .05$ | 3.4 | $1.6 \pm .24$ | 1.75 |

# Is there any Evidence for a Transport System for Glucose Derived from Sucrose in Rat Kidney?

K. Baumann and H. Vick
Max-Planck-Institut für Biophysik, Frankfurt/Main, Germany

Dr. Caspary has summarized the present knowledge (2, 3, 4, 7, 8, 9, 10, 11) of the relationships which exist between disaccharide hydrolysis and glucose transport in the small intestine. Experimental results indicate a new sodium-independent glucose transport system in hamster small intestine which is specific for glucose released by brush border disaccharidase action (2, 4, 10).

Because many similarities exist, both in structure and function, between kidney and small intestinal cells, the question arises as to the possibility of sucrose-splitting and the subsequent reabsorption of the glucose or fructose moieties in the kidney. To answer this question, proximal tubules of a rat kidney were perfused with test solutions containing free glucose or free fructose or sucrose. The sucrose was radioactively labelled at the glucose or fructose moieties. A radioactive analysis of the perfusion solution, before and after microperfusion, would reveal if the glucose or fructose moiety of sucrose disappears from the intraluminal fluid of the proximal tubules.

Male Wistar rats were maintained on a standard diet (Altromin$^R$) up to 12 hours before the experiment and tap water was always offered. Animals weighing 200g were prepared in the usual manner for tubular micropuncture (12). Proximal tubules were continuously perfused in vivo under the condition of zero net water and electrolyte flux. The perfusate c ontained per liter: sodium (120 mEq), potassium (4 mEq), calcium (3 mEq), magnesium (2 mEq), chloride (109 mEq), bicarbonate (10 mEq), acetate (10 mEq), non-reabsorbable raffinose (37 mmoles), and, in the first series of experiments, 10 mmoles D-glucose (D-glucose-1-H$^3$). Less than 1 mmol inulin (inulin-carboxyl-C$^{14}$) was always added as a marker for volume flux.

As shown in Fig. 1, free glucose readily disappeared from the intraluminal fluid under the condition of near maximum transport rates (5, 6) for glucose. The apparent $K_T$ D-glucose was measured to be 0.6 mmol/1 (5). This reabsorption could be blocked almost completely by adding 1 x $10^{-4}$mol/1 phlorizin.

In the next series of experiments proximal tubules were perfused with 10 mmol/1 sucrose (sucrose-C$^{14}$ [D-glucose-C$^{14}$(U)] ) in a glucose-free solution. The radioactivity remained almost unchanged along the perfused distance up to 3 seconds contact time (Fig. 2). The same results were obtained when the animals were kept on a low-carbohydrate diet and/or after i.p. injection of 5 ml/day of a 8.6 % sucrose solution over a period of 2 to 3 weeks. No tubular reabsorption could be detected by perfusing with 10 mmol/1 D-fructose (D-fructose-1-H$^3$) or 10 mmol/1

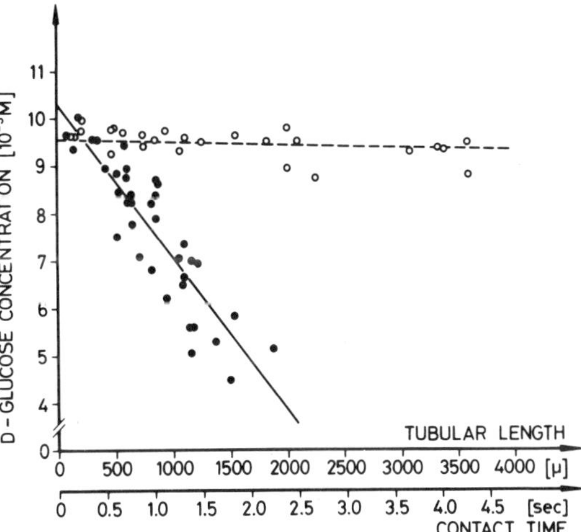

Fig. 1. Intraluminal concentration decrease of D-glucose (ordinate) as a function of perfused length of the proximal tubule or contact time (abscissa). The regression lines are calculated by Least-Square method.

●: without phlorizin
o: with $1 \times 10^{-4}$ mmol/1 phlorizin in the perfusion solution

sucrose (sucrose-H$^3$ [D-fructose-1-H$^3$(N)] ).[*] Fig. 3 shows that reabsorption of

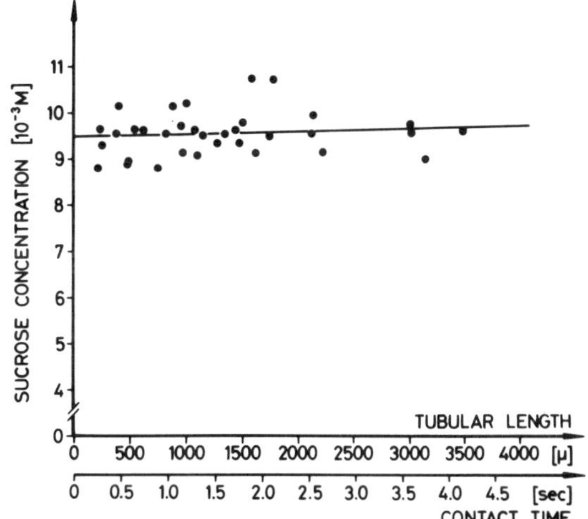

Fig. 2. Intraluminal concentration of sucrose remained almost unchanged along the proximal tubule. Ordinate and abscissa as in Fig. 1

free D-glucose (10 mmol/1) is not influenced by adding D-fructose (40 mmol/1) to the perfusion solution. The result makes the possibility of sucrose-splitting unlikely.

The failure to show a disappearance of sucrose from the intraluminal fluid of the proximal tubules of rat kidney indicates considerable differences in the function

---

[*] The radiochemical purity of all the tracer used throughout this study was checked using Whatman No. I filter paper in ascending chromatography with n-butanol-ethanol-water (52:18:30) as the solvent.

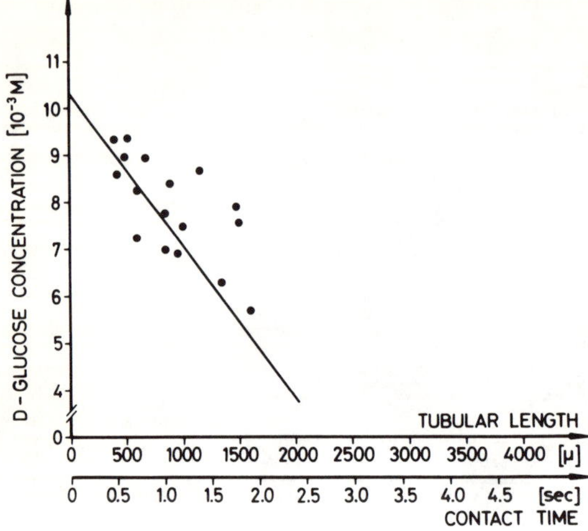



Fig. 3. Intraluminal concentration decrease of D-glucose in the presence of 40 mmol/1 D-fructose (single points) and in the absence of D-fructose (regression line from Fig. 1). Ordinate and abscissa as in Fig. 1. The points indicate that the D-glucose concentration decrease is not influenced by adding D-fructose to the perfusion solution

and structure at the molecular level and presumably in enzymatic pattern between the outer protein coat of the renal and intestinal brush borders. In agreement with our conclusion are the results of Berger and Sacktor (1) who found that sucrase is completely absent from the renal brush border or any other subcellular fraction of rabbit kidney.

## References


1. BERGER, S. J., SACKTOR, B. : Isolation and biochemical characterization of brush borders from rabbit kidney. J. Cell Biology 47, 637-645 (1970).
2. CASPARY, W. F. : Evidence for a sodium-independent transport system for glucose derived from disaccharides. (this book).
3. CRANE, R. K. : Structural and functional organization of an epithelial cell brush border. In: Intracellular transport, edited by K. B. WARREN, New York, Academic Press, 1967, Vol. V, 71-103.
4. CRANE, R. K., MALATHI, P., CASPARY, W. F., RAMASWAMY, K. : Evidence for a second glucose transport system in hamster small intestine specific for glucose released by brush border digestive enzymes. Fed. Proc. 29, 595 (1970) (abstract).
5. LOESCHKE, K., BAUMANN, K. : Kinetische Studien der D-Glucoseresorption im proximalen Konvolut der Rattenniere. Pflügers Arch. 305, 139-154 (1969).
6. LOESCHKE, K., BAUMANN, K., RENSCHLER, H., ULLRICH, K. J. ; mit einem mathematischen Anhang von FUCHS, G. : Differenzierung zwischen aktiver und passiver Komponente des D-Glucose-Transports am proximalen Konvolut der Rattenniere. Pflügers Arch. 305, 118-138 (1969).
7. MILLER, D., CRANE, R. K. : Digestive function of the epithelium of the small intestine: 1. An intracellular locus of disaccharide and sugar phosphate hydrolysis. Biochim. Biophys. Acta 52, 281-293 (1961).
8. PARSONS, D. S. : Black box models of intestinal mucosal cellular function. In: Intestinal transport of electrolytes, amino acids and sugars, W. McD. ARMSTRONG, editor, Charles THOMAS Publ., Springfield, Illinois, p. 24-51 (1971).

9. PARSONS, D. S. , PRITCHARD, J. S. : Relationship between disaccharide hydrolysis and sugar transport in amphibian small intestine. J. Physiol. 212 , 299-319 (1971).

10. RAMASWAMY, K. , MALATHI, P. , CRANE, R. K. : The direct transport of glucose from disaccharides by in vitro hamster small intestine. Fed. Proc. 30, 538 (1971) (abstract).

11. SEMENZA, G. , TOSI, R. , VALLOTON-DELACHAUX, M. C. , MÜHLHAUPT, E. : Sodium activation of human intestinal sucrase and its possible significance in the enzymatic organization of brush borders. Biochim. Biophys. Acta 89, 109-116 (1964).

12. ULLRICH, K. J. , FRÖMTER, E. , BAUMANN, K. : Micropuncture and micro-analysis in kidney physiology. In: Laboratory techniques in membrane bio-physics (H. PASSOW and R. STÄMPFLI), Springer-Verlag, Berlin-Heidel-berg-New York p. 106-129 (1969).

# A Hypothesis on the Mechanism of Mutual Inhibition among Sodium-Dependent Transport Systems in the Small Intestine


G. Semenza

Eidgenössische Technische Hochschule, Laboratorium für Biochemie, Zürich, Switzerland


During the past years a number of laboratories have reported examples of inter-actions - usually inhibitions - between different Na-dependent transport systems in the small intestine. As Dr. Alvarado (this meeting) has pointed out, examples can be found among sugars and amino acids: the net uptake of amino acids is re-duced by the presence of some sugars in the medium, and viceversa; amino acids can stimulate the net exit of sugars through the brush border membrane, and vice-versa. These surprising interactions among systems hitherto regarded as inde-pendent of each other have been the subject of considerable debate, probably for several reasons: there are quantitative species differences (as shown by Robinson and Alvarado, 1971): metabolisable sugars, even if supplied from the serosal side, increase the mucosal-serosal amino acid transfer, as shown by Smyth's group (Bingham et al., 1966); the trans-intestinal water flow induced by glucose increases the mucosal-serosal transfer of amino acids, (as shown by Munck, 1968); finally, the different experimental conditions used in the different laboratories may have led to different rate-limiting steps being measured. "At least four hypotheses have been put forward: a) inhibition due to the formation of toxic metabolites (Perry et al., 1956; Saunders and Isselbacher, 1965); b) competition for energy (Newey and Smyth, 1964; Bingham et al., 1966; Munck, 1968a, b; Reiser and Christiansen, 1969); c) stimulation of the efflux of the substrate from the cell (Chez et al., 1966; Read, 1967); and d) an inhibition between the two groups of substances at the level of the brush border membrane (Alverado, 1966; Hindmarsh et al., 1966; Casey et al., 1969), which is probably an allosteric effect due to the proximity of related binding sites in a polyfunctional matrix (Alvarado, 1966, 1968, 1970a, b, c)". (quoted from Robinson and Alvarado, 1971). Let me suggest here one possible mechanism which combines the latter three hypotheses and which can explain, in my view, the events taking place at the brush border pole of the enterocyte in a unitarian framework. I want to make clear at the onset, however, that I do not intend to discard as unlikely other mechanisms, particularly those occurring elsewhere in the cell.

Let us consider the architecture of the brush borders. It is conceivable that solutes entering the long and thin microvillus through the cell membrane would diffuse in-to the cytoplasm more slowly than they enter the microvillus. If $Na^+$ is a co-sub-strate, a local hyperconcentration of this cation will build up at the cellular face of the microvillus membrane, first near the carrier which has brought in the $Na^+$, and then near other carriers nearby. A decrease, and perhaps even an inversion of the $Na^+$ gradient results, which can have either one or both the following ef-fects: it can stimulate the exit of a substrate; it can produce a trans-inhibition of the influx of a substrate. Which of the two effects actually occurs (or perhaps both), depends on the intrinsic kinetic parameters of the system involved: for ex.: it can be shown easily that the trans-inhibition of substrate influx requires that the "permeability coefficient" (whatever its actual physical meaning) of the Na-loaded carrier should be smaller than the corresponding coefficient of the Na-free carrier. If $K^+$ is exchanged against $Na^+$ during the uptake of substrates, similar considerations can be made with regard to the $K^+$ gradient.

Clearly, this hypothesis can be regarded as a specification of the "energy competition" hypothesis of Smyth, the energy being here the $Na^+$, and perhaps $K^+$, gradients across the brush border membrane. It is not too dissimilar from Read's hypothesis (1967) and is not a rejection of Alvarado's (1966; 1968; 1970) "allosteric hypothesis".

Irrespective of whether the inhibition of net uptake occurs by _inhibition of influx_ or by _stimulation of efflux_ (both mechanisms were actually demonstrated: Alvarado, 1968; Chez et al., 1966; Robinson and Alvarado 1971) our hypothesis (which we will call the "trans-$Na^+$ hypothesis", for short) explains why:
a) mutual inhibition phenomena are only observed between substrates which are transported by $Na^+$-dependent systems (one apparent exception, the inhibition of phenylalanine net uptake by fructose in rat small intestine (Robinson and Alvarado 1971) is no longer an exception, since fructose uptake also is now known to be $Na^+$-dependent: (Gracey, et al. 1970)). This does not mean that mutual interactions between $Na^+$-dependent systems must occur in every case. Systems not closely located for ex., cannot interact in this way.
b) competitive inhibitors (even those requiring $Na^+$: phlorizin, Frasch et al., 1970; L-fucose, Caspary et al., 1969) are ineffective in eliciting heterologous inhibition of net uptake, although the actual substrates of the same transport systems do so (Alvarado, 1966; Caspary and Crane 1970).
c) sometimes the uptake of a substrate (e.g. inositol) is inhibited by heterologous substrates (sugars), although the converse is not true: The capacity of the inositol transport system is about one hundredth of that of the sugar transport system (Caspary and Crane, 1970). Therefore, sugars can bring in enough $Na^+$ to inhibit inositol uptake, but inositol will not bring in enough $Na^+$ to inhibit sugar uptake to a comparable extent. Alternatively, one of Na-dependent transport systems may have those kinetic  properties (e.g., it may have identical permeability coefficients for all forms of the carrier) which make it insensitive to a trans-inhibition of influx by $Na^+$.
d) species differences may exist, depending on a number of possible variations (intrinsic kinetic properties of the system involved, their anatomical location etc.). It should be kept in mind that interactions of this kind probably have little evolutionary advantage, if any.

These points and a number of others have been discussed elsewhere in detail (Semenza, 1971). To the best of my knowledge, the "trans-Na hypothesis" is not contradicted by any experiment published.



References

ALVARADO, F.: Transport of sugars and amino-acids in the intestine: Evidence for a common carrier. Science 151, 1010-1013 (1966).
ALVARADO, F.: Amino-acid transport in hamster small intestine: Site of inhibition by D-galactose. Nature (London) 219, 276-277 (1968).
ALVARADO, F.: Intestinal transport of sugars and amino acids - Independence or federalism? Amer. J. clin. Nutr. 23, 824-828 (1970a).
ALVARADO, F.: La membrana celular como mosaico de funciones. Bol. Soc. espan. Hist. nat. (in press) (1970b).

ALVARADO, F.: Interrelation of transport systems for sugars and amino acids in small intestine. In: Intestinal transport of electrolytes, amino acids and sugars. (Editors: W. MdD. Armstrong and A. S. Nunn), Thomas Co., Springfield, Ill.: 1970c, p. 281-318.

BINGHAM, J. K., NEWEY, H., SMYTH, D. H.: Interaction of sugars and amino acids in intestinal transfer. Biochim. biophys. Acta 130, 281-284 (1966).

CASEY, M. G., FELBER, J.-P., VANNOTTI, A.: Biochemical study of the mechanism of intestinal absorption. Amer. J. Proctol. 20, 64-67 (1969).

CASPARY, W. F., CRANE, R. K.: Active transport of myo-inosit and its relation to the sugar transport system in hamster small intestine. Biochim. biophys. Acta 203, 308 (1970).

CASPARY, W. F., STEVENSON, N. R., CRANE, R. K.: Evidence for an intermediate step in carrier-mediated sugar trans-location across the brush border membrane of hamster small intestine. Biochim. biophys. Acta 193, 168-178 (1969).

CHEZ, R. A., SCHULTZ, S. G., CURRAN, P. F.: Effect of sugar on transport of alanine in intestine. Science 153, 1012-1013 (1966).

FRASCH, W., FROHNERT, P. P., BODE, F., BAUMANN, K., KINNE, R.: Competitive Inhibition of Phlorizin Binding by D-Glucose and the influence of sodium: a study on isolated brush border membrane of rat kidney. Pflügers Archiv 320, 265-284 (1970).

GRACEY, M., BURKE, V., OSHIN, A.: Intestinal transport of fructose. Lancet ii, 827 (1970).

HINDMARSH, J. T., KILBY, D., WISEMAN, G.: Effect of amino acids on sugar absorption. J. Physiol. (London) 186, 166-174 (1966).

MUNCK, B. G.: Amino acid transport by the small intestine of the rat. Effects of glucose on the transintestinal transport of proline and valine. Biochim. biophys. Acta 150, 82-91 (1968a).

MUNCK, B. C.: Amino acid transport by the small intestine of the rat. Evidence against interactions between sugars and amino acids at the carrier level. Biochim. biophys. Acta 156, 192-194 (1968b).

NEWEY, H., SMYTH, D. H.: Effects of sugars on intestinal transfer of amino-acids. Nature (London) 202, 400-401 (1964).

PERRY, J. W., MOORE, A. E., THOMAS, D. A., HIRD, F. J. R.: Galactose intolerance: observations on the experimental animal. Acta Paediat. (Uppsala) 45, 228-240 (1956).

READ, C. P.: Studies on membrane transport. I. A common transport system for sugars and amino acids? Biol. Bull. 133, 630-642 (1967).

REISER, S., CHRISTIANSEN, P. A.: Intestinal transport of amino acids as affected by sugars. Amer. J. Physiol. 216, 915-924 (1969).

ROBINSON, J. W. L., ALVARADO, F.: Interaction between the sugar and amino-acid transport systems at the small intestinal brush border - a comparative study. Pflügers Archiv 326, 48-75 (1971).

SAUNDERS, S. J., ISSELBACHER, K. J.: Inhibition of intestinal amino acid transport by hexoses. Biochim. biophys. Acta 102, 397-409 (1965).

SEMENZA, G.: On the mechanism of mutual inhibition among sodium-dependent transport systems in the small intestine. A hypothesis. Biochim. biophys. Acta 241, 637-649 (1971).

# Does the Stoichiometry of Coupling Necessarily Reveal the Composition of the Ternary Complex?


Halvor N. Christensen
Department of Biological Chemistry, The University of Michigan, Ann Arbor,
Michigan, USA


The second subject on which I have been asked to make a summary comment is
the stoichiometry of the flux of an alkali-metal ion, on the one hand, and of an organic metabolite, on the other. Most of the results I will present depart from those
reported for System A of the Ehrlich cell, and also from those just discussed by
Dr. Curran. These departures do not, however, bring the alkali-metal ion-gradient hypothesis into question. Instead they show us that we must not always expect energy transfer to accompany linkage of fluxes, and also warn us that we must
take precautions that we are examining a homogeneous transport process.

It was K. P. Wheeler who initiated the demonstrations in our laboratory that the
flux augmentation ratios, $\Delta v_{Na}/\Delta v_{amino\ acid}$, do not necessarily correspond
to the stoichiometric composition of the ternary complex as indicated by the kinetics (1). Table I serves to summarize relations as we saw them at the termination
of Wheeler's studies in our laboratory. Note particularly the result for alanine in
the pigeon red blood cell and in the rabbit reticulocyte. The flux ratio for the latter
cell is also about 2.5. Yet in both cells the kinetics are first order with respect to
$Na^+$. Actually other substrates of the same system in these two red blood cells
show flux augmentation ratios ranging all the way from 0.2 to 5. Fig. 1 comes
from the work of Koser in my laboratory, and shows that the flux ratios for alanine and serine were not in any clear way dependent on time, nor on alanine concentration. The same was the case for $Na^+$ concentration (5).

The only important factors we recognize as producing variation in this ratio is the
structure of the external and the internal amino acid engaged in exchange. I will
illustrate this effect only for the external amino acid. Table II, also from Koser's
experiments (5), shows that introducing a hydroxyl group on carbon 3 or 4 causes
a sharp increase in the flux ratio. This effect is not seen if the hydroxyl group is
placed on carbon 5, nor if it is placed in the cis orientation on carbon 3 or carbon
4 of proline.

Fig. 2 from the work of Thomas (6) shows that the dependency of the rate of amino
acid uptake, as Wheeler showed, is first order, both with respect to the amino acid
concentration and with respect to the $Na^+$ concentration. The intercepts to the left
of the ordinate have no theoretical meaning and do not measure the dissociation of
the binary complexes, because the equilibrium assumption is not met by this system.
Perhaps I may comment that we have not yet encountered a transport system that
has proved on study to justify that assumption. The rate-limiting event here clearly is not the translocation step, or any other step common to the two substrates,


---

$^+$ The experiments from our laboratory described here have been supported in
part by a grant (HD-01233) from the Institute for Child Health and Human Development, National Institutes of Health, U.S.P.H.S.

because the rate is different for each cosubstrate.

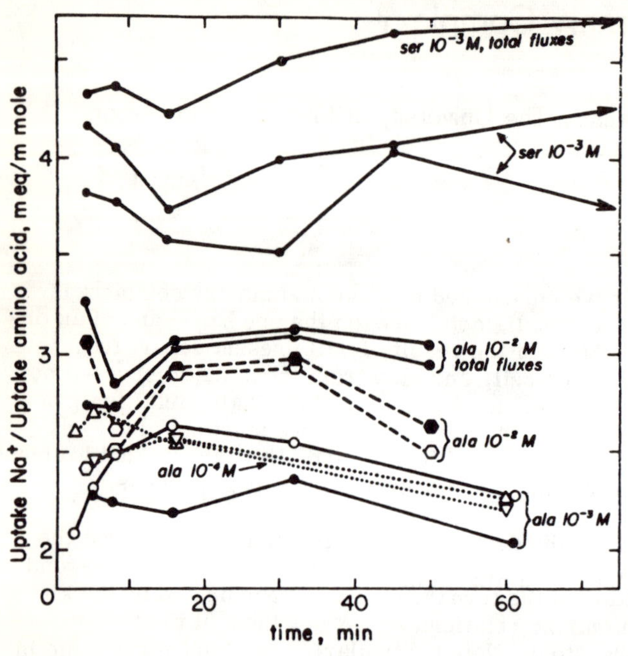



Fig. 1. Absence of a relation of the flux augmentation ratio to interval of observation for serine and alanine, also to amino acid concentration, for alanine. The acceleration of the uptake of Na$^+$ due to the presence of the indicated concentration of the amino acid, has been divided by the acceleration of the uptake of the amino acid due to the presence of Na$^+$ at 0. 01 to 0. 14 N. See reference 5  for experimental details. Reproduced with permission from Koser and Christensen, Biochim. Biophys. Acta <u>241</u>, 9 (1971)

Fig. 3 illustrates that the affinity of the amino acid is sharply increased when the hydroxyl group is in the optimal position. Note that results are shown for both 25 and 118 mN Na$^{++}$. The responsivity to the presence of Na$^+$, as indicated by the separation of the two lines, rises sharply when the hydroxyl is in the optimal position. Note that the ordinate scale for the right-hand figure has been constricted by one half. Fig. 10 of my preceding presention in this Conference has already shown diagrammatically what we think is the structural feature at the receptor site that

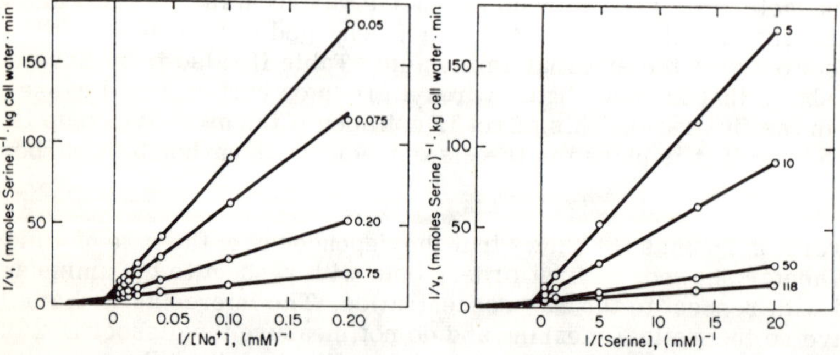



Fig. 2. Double reciprocal plots relating the rate of Na$^+$-dependent serine uptake by the pigeon red blood cell to Na$^+$ concentration (left) and to serine concentration (right). Because Na$^+$ uptake is proportional to amino acid uptake at all points, the points of intersection on the abscissa have the same values as would be given by a plot of Na$^+$ uptake against serine concentration. See reference 6 for further details. Reproduced with permission from Thomas and Christensen, J. Biol. Chem. <u>246</u>, 1682 (1971)

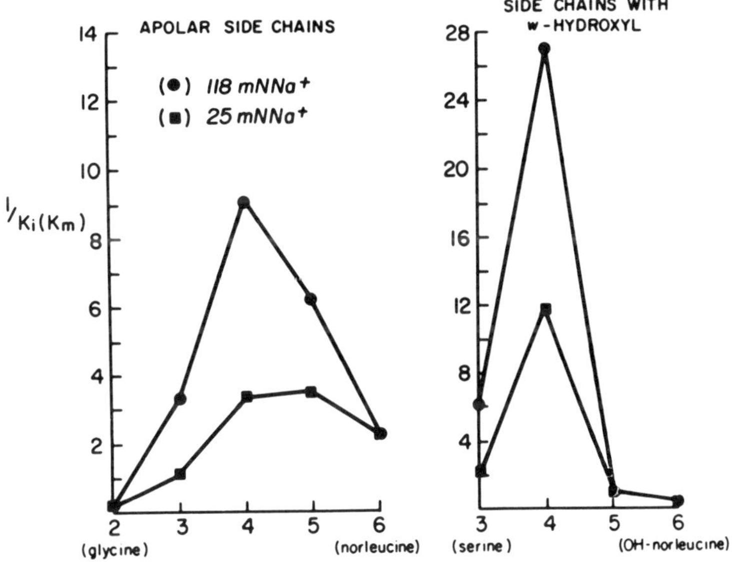



Fig. 3. Effect of chain length and position of a terminal hydroxyl group on the apparent affinity of neutral amino acids for System ASC of the pigeon red blood cell. The values shown are reciprocals of $K_i$ or $K_m$ values. Aliphatic amino acids are shown at the left, the corresponding hydroxyaliphatic amino acids at the right. Reproduced with permission from Thomas and Christensen, J. Biol. Chem. 246, 1682 (1971), which may be consulted for further details

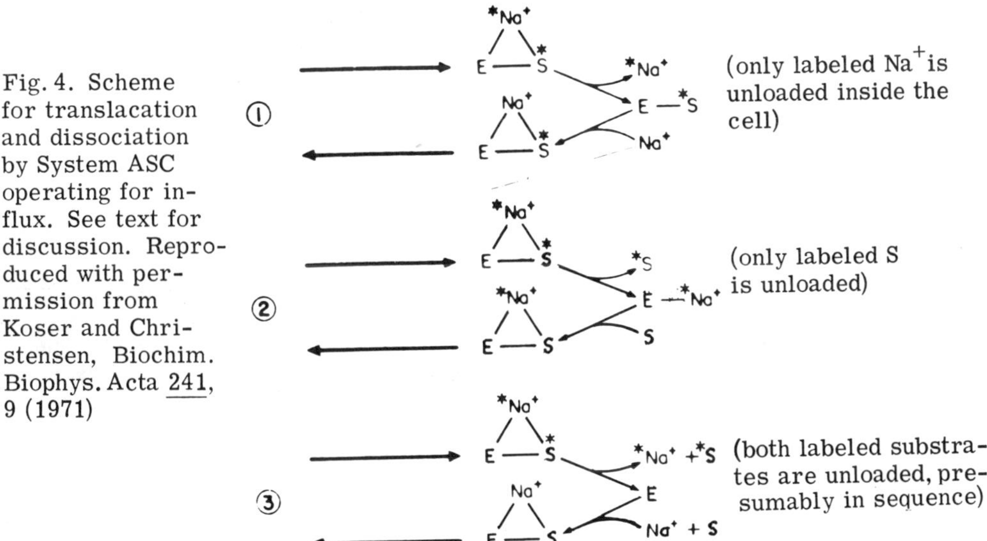



Fig. 4. Scheme for translacation and dissociation by System ASC operating for influx. See text for discussion. Reproduced with permission from Koser and Christensen, Biochim. Biophys. Acta 241, 9 (1971)

causes the interaction between the two cosubstrates, $Na^+$ and the amino acid, to be so sensitive to the position and the orientation of the hydroxyl group. We believe that the hydroxyl, mercapto or carboxamide group of the amino acid sidechain has encountered the $Na^+$ at the receptor site and formed a bond to it, to account for the sharp optimum in the position and orientation of such groups.

These researches together bring us to the mechanistic model of Fig. 4 for system ASC of the two red blood cells. First the ternary complex is formed, an event not shown here. From the results obtained by Thomas it appears that formation probably occurs in random order. The complex then oscillates from an orientation to one face of the membrane to an orientation to the other face. We are able to recognize, however, only those oscillations in which one of the cosubstrates dissociates from the complex. Either cosubstrate may be dissociated and replaced, but apparently it is comparatively uncommon that both are dissociated in the course of a given oscillation, again presumably in random order. This circumstance suggests that many oscillations occur unmarked by a dissociative step.

The translocation event obviously then is not rate limiting. Instead the dissociation-and-replacement event must be the slowest step. This step we suspect is probably a displacement reaction; i. e., a free sodium ion or amino acid molecule collides with the ternary complex and reacts in such a way as to replace the corresponding component. A change in the amino acid structure that enhances its reactivity, e. g., introduction of a properly orientated hydroxyl or sulfhydryl group, enhances $Na^+$ binding and accelerates transport of the amino acid into the cell, presumably by accelerating its release from the ternary complex; it enhances much more strongly, however, the dissociation of the sodium ion from the ternary complex, a conclusion that seems inescapable from the rise of the flux augmentation ratio.

Both substrates must have a suitable structure, and both must be present simultaneously on both sides of the membrane, transport in the form of the binary complexes being indetectible (Table III). That is, omission of one substrate leaves the uptake of the other substrate no longer sensitive to characteristic inhibitors of the system, namely arginine, homoarginine or lithium ion. Accordingly, coupling appears to be complete in the sense that the two fluxes, that of $Na^+$ and that of the amino acid, stop together; and yet it is a peculiar coupling in that several molecules of one species may migrate for one particle of the other. The possibility of complete coupling between two flows is more easily pictured as based on an integral stoichiometry corresponding to the composition of an intermediate molecular species common to the two flows. If any metabolic feature were found that could modify the flux augmentation ratio for a given amino acid with $Na^+$, the interesting possibility of a controlled relaxation of coupling in this system would be added; so far, however, no such feature is recognized.

The system described here differs sharply from the model just proposed by Doctor Curran for the uptake of alanine by the rabbit intestine. I am tempted to suggest that the alanine uptake by the intestine he sees in the absence of $Na^+$ might occur by a separate, $Na^+$-independent transport system, rather than in the binary complex, Ala-X. In that connection, Doctor Tsang-Cheng Shao and I have observed a $Na^+$-independent component of amino acid uptake by the ileum of both the rat and the hamster. On inhibition analysis this component shows a somewhat different specificity as to amino acid structure than either of the two $Na^+$-dependent transport systems observed in these two tissues.

I want to suggest that the phenomenon of obligatory exchange shown by the System ASC may not have the basis usually proposed for exchange, namely a relatively slow or forbidden movement of the empty carrier site from one orientation to the other. Instead I suspect that the rate at which the empty carrier can oscillate may not be a factor in the kinetics of this system. If each cosubstrate must be displaced by an analogous molecule or ion before recordable transport can occur, the empty carrier or the half-filled carrier would not serve to a significant extent as a reactant in the process. Dominance by displacement reactions could be a consequence of a relatively slow or absent direct dissociation of the ternary complex.

This curious case represents by no means a refutation of the alkali-metal ion-gradient hypothesis. The system cannot cause energization of amino acid transport at the steady state by the flow of $Na^+$ through the system, because $Na^+$ is exchanged for $Na^+$, and apparently does not show a net flow. But neither can the energy inherent in the several gradients be dissipated by the system, because of the tight coupling. Besides a rapid, superficially meaningless amino-acid-dependent exchange-diffusion of $Na^+$, the net result of the operation of the system appears to be a distribution of the energy inherent in the amino acid gradients among the various substrate amino acids. A factor determining the steady-state pattern of this distribution should be the value of the flux augmentation ratio characteristic of each of the several amino acid substrates. But these special feature appear to contrast with system A, which has exceedingly weak exchanging properties at least for the amino acid component, although the possibility of exchange fluxes of $Na^+$ by it may not have had sufficient consideration. It may be only one small evolutionary or differentiative change which has lost for system ASC the capacity for net energy transfer. That feature whose loss has locked the system into exchange we suppose may well be a sufficiently rapid direct dissociation of the ternary complex.

The case shows us that we must not always expect to find net energy transfer where linked fluxes are found; also that we must select our substrates carefully if our results are not to be confounded by the simultaneous participation of a system resembling ASC and one in which effective energy transfer occurs. Test in the Ehrlich cell of energization of the transport by alkali-metal-ion flows would give highly confusing results if, for example, homoserine were chosen as the test substrate. Unless sufficient care is taken to secure homogeneous transport by a single agency, heterodox indications are inevitable.

In that connection Doctor L. R. Smith and I are currently examining the flux augmentation ratios for System A of the Ehrlich ascites tumor cell. In doing so, we have taken advantage of the circumstance that $Li^+$ will serve almost as well as $Na^+$ as a cosubstrate of System A, whereas System ASC becomes imperceptible when $Li^+$ is substituted for $Na^+$ (Table III). Accordingly, by measuring the lithium-dependent uptake ($Li^+$ at 80 mN and choline at 63 mN), we limit uptake to System A and can presumably study the fluxes by that system alone. In that way the study could presumably be extended beyond glycine and AIB to amino acids that ordinarily serve simultaneously as substrates for another transport system. The net uptake of lithium by the cell during 1-min experiments was measured by atomic absorption spectrophotometry. As Table IV shows, the observed flux augmentation ratios for nine amino acids did not depart greatly from unity, a result that provides provisional support for the ion-gradient hypothesis.


## References


1. WHEELER, K. P., CHRISTENSEN, H. N.: J. Biol. Chem. <u>242</u>. 3782 (1967).
2. VIDAVER, G. A.: Biochemistry <u>3</u>, 662 (1964).
3. EDDY, A. A.: Biochem. J. <u>108</u>, 195 (1968).
4. JACQUEZ, J. A., SCHAFER, J. A.: Biochim. Biophys. Acta <u>193</u>, 368 (1969).
5. KOSER, B. H., CHRISTENSEN, H. N.: Biochim. Biophys. Acta <u>241</u>, 9 (1971).
6. THOMAS, E. L., CHRISTENSEN, H. N.: J. Biol. Chem. <u>246</u>, 1682 (1971).

## TABLE I

Contrast between the evidence for the composition of the ternary complex obtained from kinetics and from the stoichiometric ratios between fluxes.

| Substrate and cell | Order of dependency | $\Delta V_{Na}/\Delta V_s$ |
|---|---|---|
| Glycine in pigeon r. b. c. (Vidaver, 2) | $[Na^+]^2$ | $1.53 \pm 0.45$ |
| Alanine in pigeon r. b. c. (Wheeler, 1) | $[Na^+]$ | $2.52 \pm 0.34$ |
| ß-alanine in pigeon r. b. c. (Wheeler, 1) | $[Na^+]^2$ | $0.96 \pm 0.30$ |
| Glycine & AIB by System A, Ehrlich cell (3, 4) | $[Na^+]$ | $1.0$ |
| Alanine in the rabbit reticulocyte (5) | $[Na^+]^2$ | ca. $2.5$ |

## TABLE II

Effect of hydroxyl group on the flux augmentation ratio, $\Delta V_{Na^+}/\Delta V_{amino\ acid}$, for entry by System ASC of pigeon red blood cell (5).

| Unhydroxylated Amino Acid | | Hydroxylated Amino Acid | |
|---|---|---|---|
| proline | 0.2 | 4-hydroxyproline | 3.2 |
| alanine | 2.5 | serine, 4.0 (cysteine 5.0) | |
| $\alpha$-amino-n-butyric (?) | | threonine | 4.5 |

## TABLE III

Inhibitors of System ASC inhibit neither uptake of serine in absence of $Na^+$ (I), nor uptake of $Na^+$ in absence of substrate amino acids (II). Pigeon red blood cells. Choline, or choline plus homoarginine/methionine or $Li^+$ replaced $Na^+$ (6).

I. Rate of serine uptake, mmoles per kg cell water·min

| Inhibitor | Serine concentration | | | |
|---|---|---|---|---|
| | 0.2 mM | 0.5 mM | 1.0 mM | 2.5 mM |
| none | 0.008 | 0.020 | 0.044 | 0.092 |
| 10 mM arginine | 0.008 | 0.020 | 0.040 | 0.090 |
| 0.5 mM methionine | 0.006 | 0.020 | 0.042 | 0.094 |

II. Rate of $Na^+$ uptake, m eq per ka cell water·min

| $Na^+$ concentration, 50 mN | | $Na^+$ concentration, 10 mN | |
|---|---|---|---|
| Homoarginine concentration | $Na^+$ influx | $Li^+$ concentration | $Na^+$ influx |
| 0 | 0.4 | 0 | 0.12 |
| 5 | 0.5 | 25 | 0.12 |
| 10 | 0.3 | 50 | 0.11 |
| 25 | 0.4 | 75 | 0.13 |
| 50 | 0.4 | | |

## TABLE IV

Flux augmentation ratios observed for uptake of lithium ion and several amino acids by the Ehrlich ascites tumor cell. Values for $\Delta v_{Li} / \Delta v_{amino\ acid}$ during 1 min were determined as described in the text and for $Na^+$ in reference 1. Experiments by Lynton R. Smith.

| Amino Acid | $\Delta v_{Li} / \Delta v_{a.a.}$ |
|---|---|
| glycine | 0.8 |
| sarcosine | 0.6 |
| AIB | 1.3 |
| serine | 1.5 |
| homoserine | 1.1 |
| alanine | 1.4 |
| phenylalanine | 1.1 |
| proline | 1.0 |
| hydroxyproline | 1.1 |

# Specific Comment on the Paper Presented by Dr. H. N. Christensen

K. P. Wheeler
Department of Biological Chemistry, University of Michigan, Ann Arbor,
Michigan 48104, USA

A further point should be emphasized about the scheme Dr. Christensen has described to explain the observed variations of the stoichiometric ratio of linked $Na^+$ and amino acid fluxes. As pointed out previously, (Wheeler, K. P. and Christensen, H. N., J. Biol. Chem. 242, 3782 (1967)), the condition that the translocation step, whatever its molecular nature, is not rate-limiting enables us to explain all the experimental data in terms of the simplest ternary complex, NaES, without postulating the formation of complexes such as $Na_2ES$, and so on. This condition not only removes the restriction that the stoichiometric ratio of the linked $Na^+$ and amino acid fluxes must correspond with ratio present in the complex, but also can account for the various different kinetic curves that have been described. The nature of the curves formed by plotting the linked fluxes against the concentration of $Na^+$ or the amino acid depends only on the relative values of the various rate constants for the individual reactions described in the scheme, not on the composition of the complex transported, and a variety of kinetic relationships, including hyperbolic and sigmoid curves, can be predicted from the scheme. Although the translocation step is commonly considered to be rate-limiting, there is some experimental evidence from model membrane systems suggesting that the membrane/ water interphase itself is the primary barrier to passage of electrolytes and sugars through the membrane (Rosano, H. L., Duby, P. and Schulman, J. H., J. Phys. Chem. 65, 1704 (1961); LeFevre, P. G., Yung, C. Y. and Chaney, J. E., Arch. Biochem. Biophys. 126, 677 (1968)).

# A Sodium Dependent, Non-Carrier Mediated Transport of a Passive Diffusing Substance across the Intestinal Wall

G. Esposito, A. Faelli and V. Capraro
Institute of General Physiology, University of Milan, via Mangiagalli 32,
Milan, Italy

## Introduction

It is commonly accepted that sodium concentration in the bathing mucosal fluid strongly affects the transport of substances which are actively transported through the intestinal barrier (1, 2).

In previous papers (3-5) we have pointed out that the mobility coefficient of thiourea and acetamide (two substances which diffuse passively through the intestinal wall) decreases when NaCl of the incubating fluid is substituted with Tris-Cl, Choline-Cl or LiCl. As a consequence it seems that, besides a specific mechanism of sodium effect on actively transported substances, there is another unspecific effect on the resistance of the intestine to the passage of non-actively transported molecules. Several factors could account for such a decreased passive permeability, namely the shrinkage of the epithelial cell in a NaCl-substituted medium, the diminished area available for the diffusion in the lack of NaCl or, for instance, the varied Na:Ca ratio of the medium with a direct effect on the physico-chemical properties of the cell membrane. Aim of the present work is to find out which of these factors is responsible for the decreased permeability.

## Methods

The experiments were performed by recirculating the perfusion fluid through an open and everted intestinal tract (5). Sprague-Dawley albino male rats, initially weighing about 250 g, semistarved over a 15 day period (final percent weight decrease: 15-25%), were used.

Under barbituric narcosis, a tract of small intestine 10-15 cm long (jejunal tract) was removed from the animal at about 10 cm from pylorus. Each tract was everted, perfused at 28°C and gassed with a mixture of 95% $O_2$ and 5% $CO_2$. The basic incubating and perfusing fluid  was Krebs-Henseleit bicarbonate solution added, according to cases, with glucose 1.4 or 14 mM respectively. Unlabelled acetamide at the same concentration (10 mM) was added to the serosal and mucosal fluids. The composition of the solutions other than this one is reported in tables. At the end of the experiment, the emptied intestine was dried at 100°C overnight. The experiment was preceded by a 20-min preincubation period. The mobility coefficient ($\omega$) of acetamide is calculated according to the following equation (6):

$$n + v\,(1-\sigma)\,\bar{c} \ = \ \omega\ RT\,\triangle\,C$$

where n is the serosa to mucosa acetamide flux, v is the contemporaneous fluid flux from the mucosal to the serosal side, $\bar{c}$ is a mean of the concentration of acetamide in the two compartments (7), $\sigma$ is the reflection coefficient and $\triangle$ C is the mean of the concentration differences between the serosal and mucosal medium throughout the experimental period.

Even if we assume $\sigma$ =0, the term $v(1-\sigma)\bar{c}$, which describes the drag effect, is always very small in comparison with the acetamide flux (n) and the conclusions drawn from the "$\omega$" values calculated by assuming $\sigma$ =0 do not vary significantly from those obtained from "$\omega$" values calculated by disregarding the drag effect and following the formula:

$$n = \omega \ RT \ \Delta C$$

Therefore, in the text only these latter "$\omega$" values are reported. Net glucose transport and "$\omega$" values were at first calculated for each of all 10-20 min experimental periods and then mediated for 1 h experiment. The intracellular water and the acetamide concentration in the epithelial cell, as well as the partial concentration differences between serosa and cell and between cell and mucosa (see Table 4), were determined after 30-min experiments using tritiated inulin or [14]C-polyethylenglycol. For further details see ref. 5.


Results and Discussion

It has to be pointed out that the substance here employed (acetamide) crosses the intestinal wall according to a simple thermal diffusion,as seen in Table 1. This Table shows that by increasing the concentration of acetamide in the incubating media, the mobility coefficient of this substance does not statistically vary,at least in the  range of concentrations used.

Table 2 shows that in NaCl-substituted medium, the epithelial cell shrinks; as a matter of fact, the intracellular water decreases in comparison with the control experiments (P< 0.001) as well as the mobility coefficient (P< 0.001). This shrinkage could explain the lowering of the mobility coefficient of acetamide in the NaCl-free medium, but by incubating the intestine in the basic solution (glucose 14 mM) added with phlorizin $10^{-4}$M, the intestinal cell shrinks as in the NaCl-substituted solution, nevertheless the mobility coefficient is unaffected (P> 0.2). Therefore, the volume change of the epithelial cell does not account for the variation of the mobility coefficient.

It is known that during transport activities the intercellular spaces of the epithelium open up (8, 9); this phenomenon could affect the transintestinal passive permeability because of an increase of the total serosal surface available for the diffusion. However, if we compare (Table 3) a group of intestinal tracts perfused with a low glucose content (1.4 mM) to a group perfused with a NaCl-substituted solution (glucose 14 mM), there  is the same fluid and glucose transport (and presumably the same degree of opening of the  intercellular spaces), nevertheless the mobility coefficient is higher in the intestine perfused with a low glucose and normal NaCl concentration. This fact seems to disprove the influence of the area on the mobility coefficient of the acetamide.

Finally, Table 4 shows that,in the NaCl-free medium, mainly the concentration difference between the cell and the mucosal side is increased; therefore the resistance of the brush border to the diffusion of the tested substance seems to be much more affected than the resistance at the serosal one.

It seems reasonable to conclude that the lowering of sodium concentration in the incubating medium could directly affect the physico-chemical properties of the cell membrane (brush border) due, for instance, to the varied Na:Ca ratio of the solution and consequently the passive permeability of a hydrosoluble, uncharged and non-metabolized substance which passively permeates the intestinal wall.

## Summary

The passive permeability of acetamide through the jejunal tract of rat intestine perfused "in vitro" is significantly decreased by lowering the sodium concentration in the perfusing fluids. Sodium seems to affect directly the physico-chemical properties of the membranes (brush border) of the epithelial cell.

## Key words

Sodium and intestinal passive permeability - Acetamide - Mobility coefficient.

## References

1. CSAKY, T. Z. , THALE, M. : J. Physiol. 151, 59 (1960).
2. CRANE, R. K. : Fed. Proc. 21, 891 (1962).
3. ESPOSITO, G. , FAELLI, A. , CAPRARO, V. : Experientia 25, 603 (1969).
4. FAELLI, A. , ESPOSITO, G. , GAROTTA, G. , CAPRARO, V. : Experientia 27, 652 (1971).
5. ESPOSITO, G. , FAELLI, A. , CAPRARO, V. : "Permeability and Function of Biological Membranes" (ed. L. Bolis, North-Holland Publ. Co. , Amsterdam 1970), p. 74.
6. KEDEM, O. , KATCHALSKY, A. : Biochim. Biophys. Acta 27, 229 (1958).
7. LIFSON, N. , HAKIM, A. A. : Am. J. Physiol. 211, 1137 (1966).
8. TORMEY, J.McD. , DIAMOND, J. M. : J. Gen. Physiol. 50, 2031 (1967).
9. LOESCHKE, K. , BENTZEL, C. J. , CSAKY, T. Z. : Am. J. Physiol. 218, 1723 (1970).

**TABLE 1**

The incubating fluid is a Krebs-Henseleit bicarbonate solution added with glucose 14 mM and acetamide. Mean values $\pm$ S. E. M., referred to 1 g of dry tissue weight and 1 h, are reported. Number of experiments in parentheses.

| Concentration of acetamide in the incubating fluids | | Mobility $(\omega)$ of acetamide $\mu$ moles$\cdot$g$^{-1}\cdot$h$^{-1}\cdot$atm$^{-1}$ |
|---|---|---|
| 1 mM | (4) | 634 $\pm$ 97 |
| 10 mM | (14) | 721 $\pm$ 55 |
| 20 mM | (4) | 689 $\pm$ 35 |

TABLE 2

Intracellular water content, mobility of acetamide ($\omega$) and net glucose transport across rat jejunum. Mean values $\pm$ S. E. M. referred to 1 g dry tissue weight and 1 h, are reported. Number of experiments in parentheses.

| Incubation fluid | Intracellular water content $ml \cdot g^{-1}$ | "$\omega$" $\mu moles \cdot g^{-1} \cdot h^{-1} \cdot atm^{-1}$ | Net glucose transport $\mu moles \cdot g^{-1} \cdot h^{-1}$ |
|---|---|---|---|
| Basic perfusion fluid + glucose 14 mM | $5.56 \pm 0.15$ (9) | $721 \pm 55$ (14) | $280 \pm 51$ (14) |
| Phlorizin added $10^{-4}$ M | $4.50 \pm 0.17$ (8) | $634 \pm 57$ (5) | $-14 \pm 12$ (5) |
| Tris-Cl substituted isosmotically for NaCl | $4.60 \pm 0.14$ (12) | $354 \pm 34$ (9) | $75 \pm 17$ (9) |

## TABLE 3

Net glucose and fluid transport and mobility of acetamide ($\omega$) across rat jejunum. Mean values $\pm$ S. E. M. obtained when the range of glucose transport is between 0 and 100 $\mu$moles$\cdot$g$^{-1}\cdot$h$^{-1}$. All data are referred to 1 g dry tissue weight and 1 h. Number of experiments in parentheses.

| Incubation fluid | Net glucose transport $\mu$moles$\cdot$g$^{-1}\cdot$h$^{-1}$ | Net fluid transport ml$\cdot$g$^{-1}\cdot$h$^{-1}$ | "$\omega$" $\mu$moles$\cdot$g$^{-1}\cdot$h$^{-1}\cdot$atm$^{-1}$ |
|---|---|---|---|
| Basic perfusion fluid + glucose 1.4 or 14 mM | $35 \pm 9$ (10) | $2.59 \pm 0.25$ (10) | $439 \pm 24$ (10) |
| Tris-Cl substituted isosmitically for NaCl + glucose 14 mM | $46 \pm 14$ (6) | $2.75 \pm 0.74$ (6) | $315 \pm 30$ (6) |
| | (P > 0.4) | (P > 0.7) | (P < 0.01) |

TABLE 4

Acetamide concentration difference ($\Delta$ C) between serosal and cellular space and between cellular and mucosal space and acetamide unidirectional flux (from serosal to mucosal side). Mean values $\pm$ S. E. M., referred to 1 g dry tissue weight and 1 h, are reported. Number of experiments in parentheses.

| Incubation fluids | Unidirectional flux $\mu$moles$\cdot$g$^{-1}\cdot$h$^{-1}$ | $\Delta$C between ser. and cell (mM) | $\Delta$C between cell and muc. (mM) |
|---|---|---|---|
| Basic perfusion fluids + glucose 14 mM | 85.5 $\pm$ 9.8 (6) | 4.84 $\pm$ 0.17 (8) | 0.87 $\pm$ 0.09 (8) |
| Tris-Cl substituted isosmotically for NaCl + glucose 14 mM | 68.7 $\pm$ 5.4 (8) | 5.37 $\pm$ 0.21 (8) | 1.63 $\pm$ 0.04 (8) |
| | (P > 0.1) | (P > 0.1) | (P < 0.001) |

# K$^+$-Ions, Swelling, and Sugar Transport in Muscle

Torben Clausen and Peter George Kohn
Institute of Physiology, University of Århus, 8000 Århus C, Denmark

## Introduction

In muscle cells and other insulin-sensitive tissues, both the transport and the
metabolism of glucose is markedly influenced by electrolytes (1-6). However, it
has not yet been possible to propose a convincing model which could account for
these phenomena as being due to a direct coupling between the transport of glucose
and some electrolyte. On the other hand, the work done till so far has suggested
a number of more direct relationships between Na$^+$, K$^+$ and sugar transport (5, 7, 8).

The present report describes data which may serve as an example of such indirect
relationships. It is well known that K$^+$ ions inhibit carrier-mediated sugar trans-

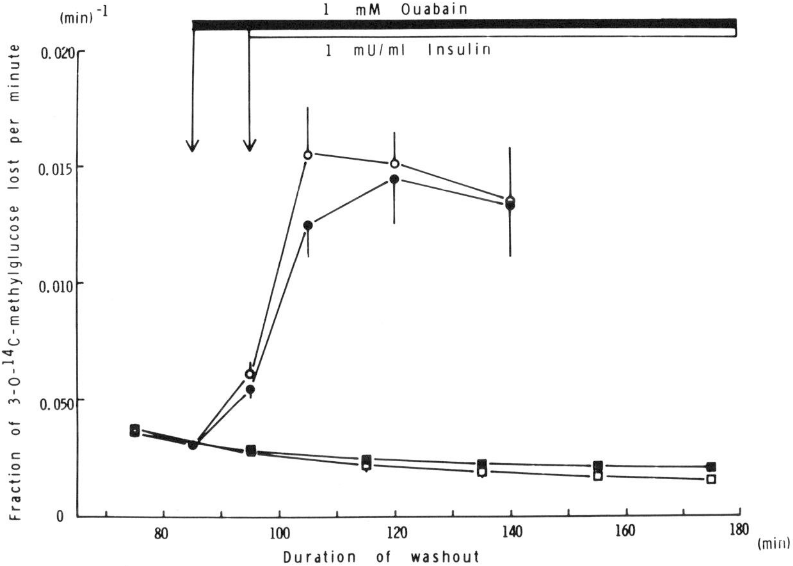


Fig. 1. Effect of ouabain and insulin on the rate coefficient for 3-0-methylglucose
release from soleus muscle. Isolated rat soleus muscles were loaded in Krebs-
Ringer bicarbonate buffer containing 1 mM pyruvate, 1 mM 3-0-methylglucose
and 3-0-$^{14}$C-methylglucose (1 or 2 $\mu$C/ml) for 60 min at 30$^\circ$. They were then
washed out in a series of tubes containing buffer with pyruvate (1 mM), but no
3-0-methylglucose. The fraction of $^{14}$C activity lost per min is shown as a function of the washout time. □ — □ , controls; ■ — ■ , ouabain (1 mM); O — O ,
insulin (0.5 mU/ml); ● — ● , ouabain (1 mM) + insulin (0.5 mU/ml). Each
point represents the mean of 3 observations with bars indicating S. E. when this
value exceeds the size of the symbols

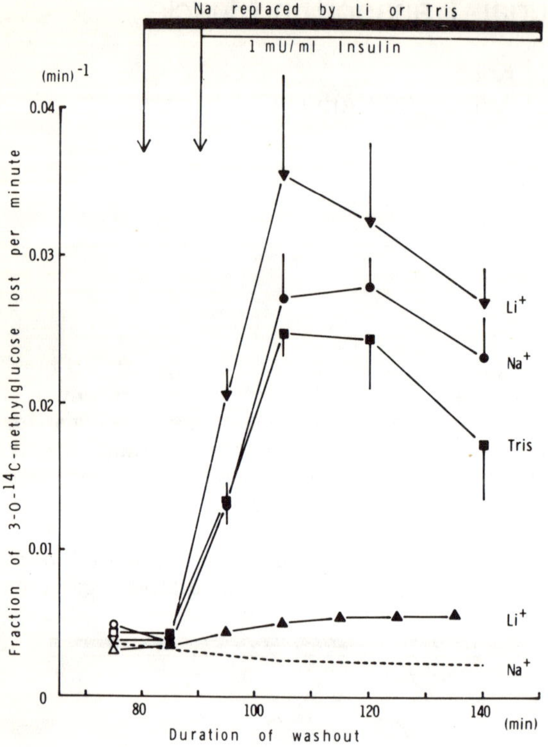



Fig. 2. Effect of $Na^+$ replacement on basal and insulin-stimulated 3-0-methylglucose release. Experimental conditions as in Fig. 1 ------, controls; ●-◉, normal Krebs-Ringer bicarbonate buffer with insulin (1 mU/ml); ▲—▲ , all $Na^+$ replaced by $Li^+$ (control); ▼—▼ , all $Na^+$ replaced by $Li^+$, insulin (1 mU/ml; ■—■ , all $Na^+$ replaced by Tris, insulin (1 mU/ml)

port, both in systems capable of performing uphill transport, and in muscle, where no such accumulation takes place. This effect has been characterized in some detail by measuring the transport of the non-metabolized sugar 3-0-methylglucose in a small intact mammalian muscle.

The results indicate that the inhibitory effect of $K^+$ ions on sugar transport is not necessarily exerted directly on the sugar transport system, but perhaps, related to the marked changes in cell volume induced by changes in the extracellular $K^+$ concentration.

## Methods

The experiments described in this communication were all performed using intact soleus muscles (25-40 mg) from fed Wistar rats (60-70 g). The methods used for the preparation of muscles, incubation, measurement of 3-0-methylglucose exchange, inulin space or electrolyte contents have been described elsewhere (8, 9, 10). The details of experimental procedures are given in the legends to figures.

## Results and Discussion

In soleus muscle, 3-0-methylglucose is transported by an equilibrating system which shares a number of characteristics with that mediating the uptake of glucose (8). Since measurements of 3-0-methylglucose uptake can only yield information

Fig. 3. Effect of K$^+$ and insulin on 3-0-methylglucose release. Experimental conditions as in Fig. 1, but 2-5 observations for each point. All muscles given insulin (1 mU/ml) from 90 min.

O—O , normal Krebs-Ringer bicarbonate buffer; ●—● , all Na$^+$ replaced by K$^+$

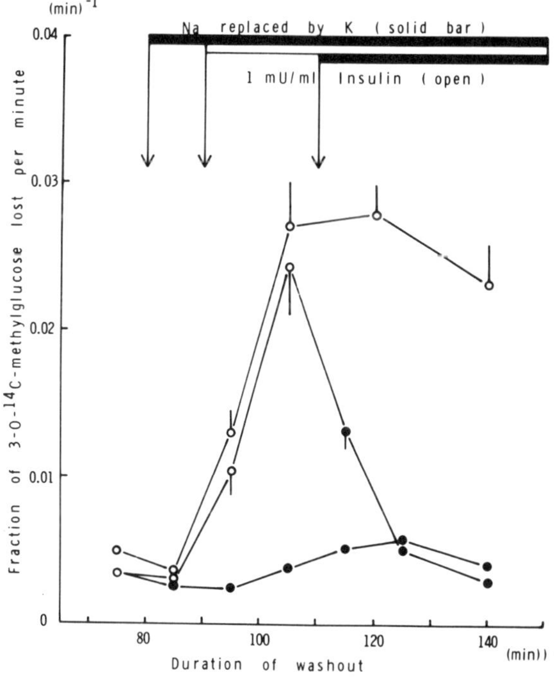


about the net result of influx and efflux, the permeability to this sugar has been evaluated by following the efflux of $^{14}$C-3-0-methylglucose from preloaded muscles into a sugar-free medium with a volume around 200 times larger than that of the intracellular space. By this procedure it is possible to measure the transport in mainly one direction and to obtain a very sensitive parameter for the permeability to 3-0-methylglucose, which at the same time describes the time course of stimulation or inhibition.

Fig. 1 shows the effect of insulin and of ouabain on the efflux of 3-0-methylglucose. Under basal conditions, the rate coefficient of 3-0-methylglucose efflux reaches a rather stable level around 60 min after the onset of washout. It can be seen that insulin (1 mU/ml), when added to the efflux medium, produces a prompt and marked rise in rate coefficient. The fact that this effect can be abolished by phlorizin or cooling indicates that the hormone activates a sugar transport system in the plasma membrane (8). On the other hand, ouabain (1 mM) produced a barely significant increase only after 90 min. Other experiments showed that, at this concentration, the glycoside suppressed $^{22}$Na efflux by 80% within 10 min. In experiments of 60 min duration, ouabain (1 mM) had no effect on the uptake of 3-0-methylglucose (8). These results correspond to earlier evidence (2, 3), and the late stimulating effect of ouabain (which seems to be more marked in the isolated rat diaphragm (11)) might be related to the contractures, which gradually develop in muscles exposed to this compound for longer periods of time. Furthermore, since ouabain does not affect the response to insulin, it seems reasonable to conclude that blocking of the active Na$^+$-K$^+$ transport is of no direct consequence for the processes of sugar transport or their activation by insulin in muscle.

Fig. 2 shows the results of experiments in which the effect of insulin (1 mU/ml) was tested in the presence and in the absence of Na$^+$. The omission of Na$^+$ from the ex-

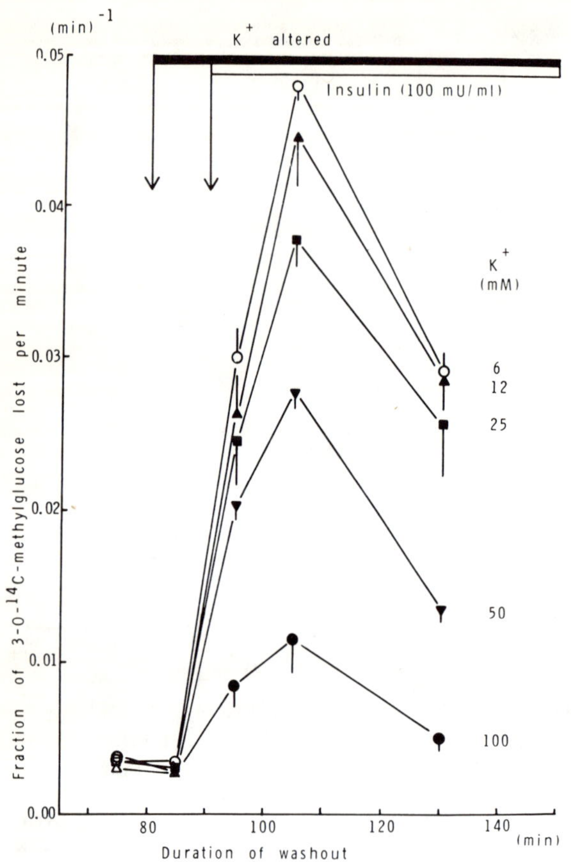



Fig. 4. Effect of varying $K^+$ levels on the stimulation of 3-0-methylglucose release by 100 mU/ml insulin. Experimental conditions as in Fig. 1. Normal Krebs-Ringer bicarbonate buffer for the first 80 min of washout, and then replacement of the indicated amount of $Na^+$ by $K^+$. Each point represents the mean of 3-10 observations

tracellular milieu caused no significant decrease in basal or insulin-stimulated 3-0-methylglucose efflux. In the $Li^+$-substituted buffer the efflux is even stimulated - an observation otherwise in agreement with earlier studies performed with diaphragm muscle (12, 13).

On the other hand, when $K^+$ is used for the substitution of $Na^+$, the stimulating effect of insulin on 3-0-methylglucose efflux is promptly suppressed to levels not far from the basal (Fig. 3). $K^+$-ions inhibit the insulin-stimulated 3-0-methylglucose efflux (even at very high concentrations of the hormone) also when present at lower concentrations (12-100 mM) (Fig. 4).

Also when 3-0-methylglucose efflux is stimulated by other means (trypsin, 2, 4-dinitrophenol, electrical stimulation), an increase in the $K^+$ concentration of the efflux medium leads to a suppression of the rate coefficient (Ref. 10 and Fig. 5, which shows the results of experiments with trypsin). However, the basal rate of efflux is not affected (Fig. 5). These results indicate that $K^+$ substitution interferes with the mechanism by which sugar transport can be accelerated, not only by insulin, but also by a less specific variety of other factors.

From Fig. 6 it can be seen that when 3-0-methylglucose transport is evaluated by measurements of the uptake, $K^+$ substitution suppresses the insulin-stimulated, but

Fig. 5. Effect of K⁺ on the stimulation of 3-0-methylglucose release by trypsin. Experimental conditions as in Fig. 1.

$\triangle$ — $\triangle$ , trypsin (0.5 mg/ml) in Krebs-Ringer bicarbonate buffer; $\bullet$ — $\bullet$ , trypsin (0.5 mg/ml) in buffer with 25 mM Na⁺ replaced by K⁺; $\blacksquare$ — $\blacksquare$ , trypsin (0.5 mg/ml) in high K⁺ buffer (100 mM Na⁺ replaced by K⁺); - - - - - - high K⁺ (100 mM) buffer without trypsin. Each point represents the mean of 3 observations

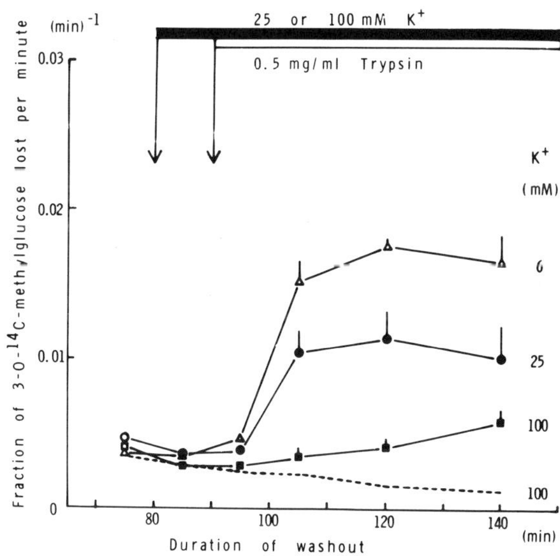


Fig. 6. Effect of Na⁺ replacement and insulin on the uptake of 3-0-methylglucose by rat soleus muscle. Soleus muscles were incubated for 60 min at 30° in 2 ml of Krebs-Ringer bicarbonate buffer containing 1 mM pyruvate, 1 mM 3-0-¹⁴C methylglucose (0.2 μC/ml) without (C) and with (I) insulin (1 mU/ml). Li⁺ and K⁺ K.R. were prepared by replacing all Na⁺ by an equivalent amount of Li⁺ or K⁺ respectively. In one experiment, only 100 mM Na⁺ was replaced by K⁺. The hatched part of the columns indicate the uptake of 3-0-methylglucose (as μ moles per g wet wt. of muscle) into the space not available to inulin. The column heights represent the values from (N) observations and 2xS.E. is denoted by the vertical bars

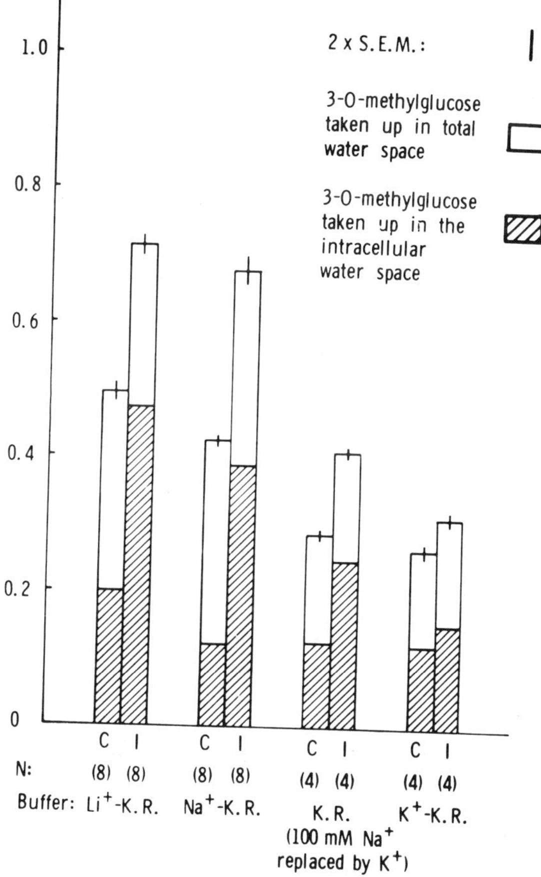

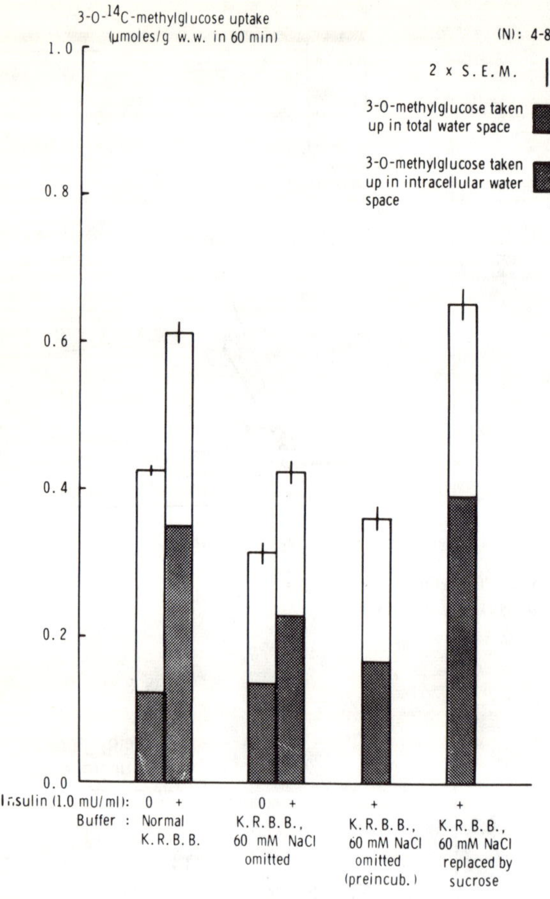



Fig. 7. Effect of hypotonicity and insulin on the uptake of 3-0-methylglucose by rat soleus muscle. Experimental conditions as in Fig. 6. The tonicity of the incubation medium was lowered by omitting 60 mM NaCl. In one experiment the muscles were preincubated for 15 min in this hypotonic buffer before the actual measurement of 3-0-methylglucose uptake. The last column shows the results of an experiment in which the osmolarity was brought back to normal by adding 120 mM sucrose to the hypotonic milieu

not the basal intracellular filling. Again, the total replacement of $Na^+$ by $Li^+$ produces no significant decrease in basal or insulin-stimulated transport into the intracellular space. This together with the data of Fig. 2 indicates that the presence of $Na^+$ in the extracellular milieu is not essential for the processes of 3-0-methylglucose transport, and, at the same time argues against the possibility that the inhibitory effect of $K^+$ substitution is the outcome of $Na^+$ lack.

It is well established that $K^+$ substitution produces swelling of muscle cells (10, 14, 15). From Fig. 6 it can be seen that in $K^+$-rich media, the extracellular (inulin) space of soleus muscle is almost halved. This observation suggested a number of experiments designed to explore the possibility that changes in cell volume might account for the inhibitory effect of $K^+$ substitution.

Fig. 7 shows that when soleus muscles were incubated in a hypotonic buffer which produced the same degree of swelling (evaluated by measurements of inulin space and water content) as the buffer in which 100 mM of $Na^+$ was replaced by $K^+$, the stimulating effect of insulin (1 mU/ml) on 3-0-methylglucose uptake is considerably suppressed. As in the $K^+$-rich medium, the uptake measured in the absence of insulin is not significantly suppressed. Since the effect of insulin was unaffected by replacing 60 mM of NaCl by sucrose, it seems unlikely that the inhibitory effect

Fig. 8. Effect of hypotonicity on basal and insulin-stimulated release of 3-0-methylglucose. Experimental conditions as in Fig. 1.

o—o , insulin (1 mU/ml) in Krebs-Ringer bicarbonate buffer; ●—● , insulin (1 mU/ml) in hypotonic buffer (60 mM NaCl omitted); □—□ , hypotonic buffer (60 mM NaCl omitted) without insulin. Each point represents the mean of 3 observations

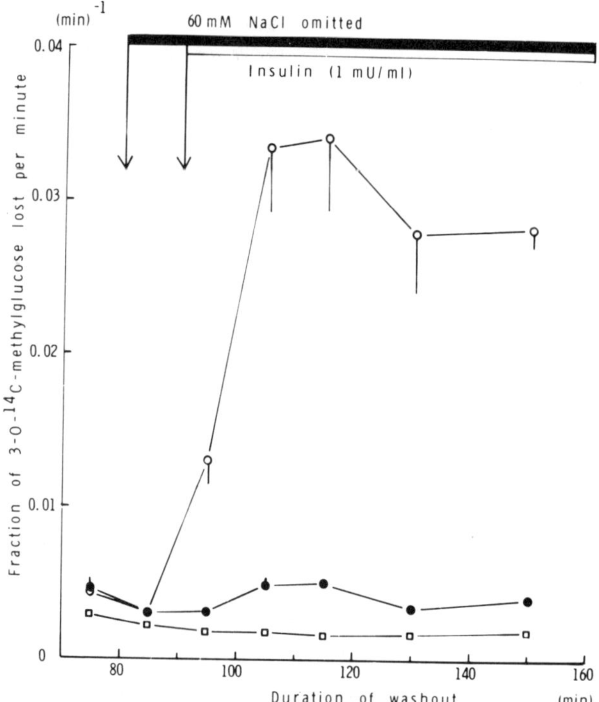


of the hypotonic milieu could be secondary to lack of sodium or chloride ions.

Washout experiments showed that hypotonicity had a similar effect on the efflux of 3-0-methylglucose. Whereas the omission of 60 mM of NaCl from the efflux medium caused no change in the basal rate of efflux, the acceleration produced by insulin (1 mU/ml) was almost abolished (Fig. 8). As was found to be the case for $K^+$ substitution, the inhibitory effect of hypotonicity was not restricted to the insulin-stimulated situation. Fig. 9 shows that the increase in rate coefficient induced by 2, 4-dinitrophenol or trypsin are both considerably suppressed when the efflux is taking place in a hypotonic environment.

These observations suggested that swelling of the muscle cells might account for a considerable part of the inhibitory effect of $K^+$ substitution. Consequently, if the swelling effect of $K^+$ could somehow be prevented, the transport of 3-0-methylglucose might not be affected by this ion. Studies with other muscles have shown that the swelling effect of $K^+$ is not seen if it is added in excess of the other components of the incubation medium (15, 16). Also in soleus muscle, it was possible to diminish the swelling if $K^+$ was added in this way, i. e. without changing the $Na^+$ concentration (Table I).

From Fig. 10 it can be seen that when 100 mM KCl is added as dry salt to the incubation medium, the rate coefficient of 3-0-methylglucose efflux is slightly increased. In this buffer, the stimulating effect of insulin is not significantly different from the obtained in the normal buffer. In contrast, when the same concentration of KCl is added as replacement for an equivalent amount of NaCl, the response to insulin (1 mU/ml) is virtually abolished (10).

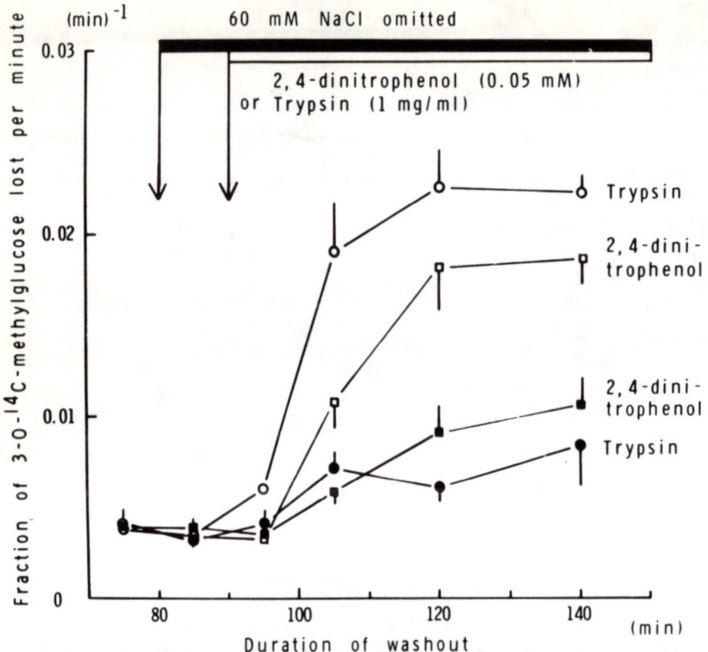



Fig. 9. Effect of hypotonicity on the stimulation of 3-0-methylglucose release by trypsin or 2, 4-dinitrophenol. Experimental conditions as in Fig. 1. o——o , trypsin (1 mg/ml) in normal Krebs-Ringer bicarbonate buffer; ●——● , trypsin (1 mg/ml) in hypotonic buffer (60 mM NaCl omitted); □——□, 2, 4-dinitrophenol (0.05 mM) in normal Krebs-Ringer bicarbonate buffer; ■——■ , 2, 4-dintrophenol (0.05 mM) in hypoosmolar buffer (60 mM NaCl omitted). Each point represents the mean of 3 observations

A separate series of experiments showed that in rat erythrocytes, the uptake of 3-0-methylglucose shows saturation kinetics and is inhibited by glucose and phlorizin. In this cell, both influx and efflux of 3-0-methylglucose is completely unaffected by changes in the extracellular $K^+$ concentration in the range 6-100 mM[x].

These data indicate that the inhibitory effect of $K^+$ is not due to a direct effect of this ion per se on the sugar transport system, but perhaps rather secondary to concomitant changes in cell volume.

The significance of extracellular osmolarity or cell volume in controlling sugar transport has repeatedly been demonstrated in studies with diaphragm muscle, adipocytes and soleus muscle (17-19, 8-10). The addition of mannitol, sorbitol, sucrose, NaCl, LiCl or choline chloride in excess of the other components of the incubation medium leads to a rapid increase in the transport of glucose or 3-0-methylglucose. On the other hand, similar concentrations of urea (9) or KCl (which are virtually without effect on cell volume) produce only a minor stimulation of sugar transport. Thus the effects of extracellular osmolarity on sugar permeability might require that the overall configuration of the cell membrane is changed by swelling or shrink-

[x] H. Harving and T. Clausen, unpublished experiments.

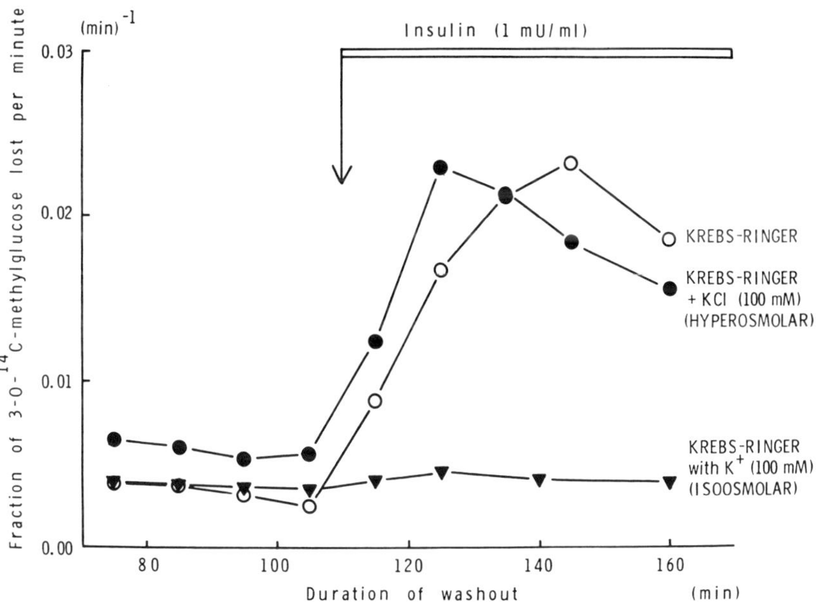



Fig. 10. Effect of $K^+$ substitution and of KCl addition on insulin-stimulated release of 3-0-methylglucose. Experimental conditions as in Fig. 1. o — o , normal Krebs-Ringer bicarbonate buffer; ● — ● , Krebs-Ringer bicarbonate buffer with the addition of KCl (100 mM); ▼ — ▼ , Krebs-Ringer bicarbonate buffer in which 100 mM of the $Na^+$ had been replaced by the equivalent amount of $K^+$. All muscles given insulin (1 mU/ml) from 90 min. Each point represents the mean of from 2-3 observations

ing. Little is known about how changes in plasma membrane area will affect its molecular structure, and the function of the sugar transport system. At present, another interpretation seems possible and more accessible to experimental testing. Several studies have shown that the transverse tubular system of muscle is readily accessible to extracellular markers (peroxidase, ferritin or labelled albumin). Consequently, the walls of these tubules (which have an area 5-7 times larger than that of the sarcolemma (20-21)) constitute a significant part of the surface across which solutes may be exchanged between the sarcoplasm and the extracellular environment. In fact, the membrane lining these tubules might mediate the major part of the sugar transport into and out of the sarcoplasm. The close association of the T-tubules with the considerably more extensive system of longitudinal sarcoplasmic tubules suggests that even larger areas of membranes might be involved in the exchange between sarcoplasm and the classical extracellular space. These structures have repeatedly been shown to undergo marked changes in volume when the muscle cell is exposed to changes in extracellular osmolarity (22-24). Shrinking or swelling may diminish or augment the accessibility of the membranes lining the tubules, and thus be decisive for the area of membrane available for transport of sugars.

On the basis of the data described above, it seems reasonable to conclude that in soleus muscle the transport of 3-0-methylglucose (and presumably that of glucose, also) is mediated by a system that is not directly linked to the processes of active $Na^+$-$K^+$ transport, and which is rather insensitive to $Na^+$. The inhibitory effect of $K^+$ on the glucose transport system in muscle resembles that described for intestine or kidney, but seems to be indirect and much less specific. The present evidence indicates that it may be accounted for as the outcome of swelling of the muscle cells.

For several years, electron microscopy has provided us with details of a complex arrangement of membranes available for solute exchange. This structural evidence should be taken into consideration as a possible basis for regulation of transport. Changes in apparent $K_M$ and $V_{max}$ of sugar transport might in some cases not reflect modifications of the carrier system per se, but rather be the outcome of changes in the relative participation of different membrane areas in the processes of exchange between cytoplasm and extracellular environment.

References


1. BHATTACHARYA, G.: Biochem. J. 79, 369 (1961).
2. KIPNIS, D.M., PARRISH, J.E.: Fed. Proc. 24, 1051 (1965).
3. CLAUSEN, T.: Biochim. Biophys Acta 109, 164 (1965).
4. HO, R.J., JEANRENAUD, B.: Biochim. Biophys. Acta 144, 61 (1967).
5. CLAUSEN, T.: Biochim. Biophys. Acta 183, 625 (1969).
6. BIHLER, I., SAWH, P.C., ELBRINK, J.: Fed. Proc. 30, 2, 256 (1971).
7. CLAUSEN, T.: Hormone and Metabolic Res., Suppl. 2, p. 66 (1970).
8. KOHN, P.G., CLAUSEN, T.: Biochim. Biophys. Acta 225, 277 (1971).
9. CLAUSEN, T., GLIEMANN, J., VINTEN, J., KOHN, P.G.: Biochim. Biophys. Acta 211, 233 (1970).
10. KOHN, P.G., CLAUSEN, T.: Biochim. Biophys. Acta 1971 (in press)
11. BIHLER, I.: Biochim. Biophys. Acta 163, 401 (1968).
12. BHATTACHARYA, G.: Biochim. Biophys. Acta 93, 644 (1964).
13. CLAUSEN, T.: Biochim. Biophys. Acta 150, 66 (1968).
14. BOYLE, P.J., CONWAY, E.J.: J. Physiol. 100, 1 (1941).
15. REUBEN, J.P., GIRARDIER, L., GRUNDFEST, H.: J. Gen. Physiol. 47, 1141 (1964).
16. GAINER, H., GRUNDFEST, H.: J. Gen. Physiol. 51, 399 (1968).
17. KUZUYA, T., SAMOLS, E., WILLIAMS, R.H.: J. Biol. Chem. 240, 2277 (1965).
18. CLAUSEN, T.: Biochim. Biophys. Acta 150, 56 (1968).
19. GLIEMANN, J.: Diabetes 14, 643 (1965).
20. PEACHEY, L.D.: J. Cell Biol. 25, No. 3, Part 2, 209 (1965).
21. FALK, G., FATT, P.: Proc. Roy. Soc. London, Ser. B 160, 69 (1964).
22. HUXLEY, H.E., PAGE, S., WILKIE, D.R.: J. Physiol. London 169, 325 (1963).
23. GIRARDIER, L., REUBEN, J.P., BRANDT, P.W., GRUNDFEST, H.: J. Gen. Physiol. 47, 189 (1963).
24. FREYGANG, W.H., jr., GOLDSTEIN, D.A., HELLAM, D.C., PEACHEY, L.D. J. Gen. Physiol. 48, 235 (1964).

## TABLE I

Effect of $K^+$ on wet weight, inulin space and intracellular water space in rat muscle.

Soleus muscles were incubated for 60 min in normal or modified Krebs-Ringer bicarbonate buffer containing 0.1 $\mu$C/ml of $^{14}$C-labelled inulin. The muscles were weighed before and after incubation, and the inulin space estimated from the $^{14}$C-activity of trichloracetic acid (5%) extracts of the muscles (ref. 8). The results are given as mean $\pm$ S.E.M. with the number of observations in parantheses.

| Buffer | | wet weight of muscles (mg) | | Inulin-space (% of w.w.) | | Per cent change in intracellular water space during incubation | Significance of difference between control exptl. |
|---|---|---|---|---|---|---|---|
| Krebs-Ringer bicarbonate buffer | Before incubation: | 35.4 $\pm$ 1.0 | (4) | | | | |
| | After incubation: | 35.8 $\pm$ 1.3 | (4) | 28.3 $\pm$ 1.0 | (4) | + 0.4 | |
| Krebs-Ringer bicarbonate buffer + 100 mM KCl | Before incubation: | 35.9 $\pm$ 0.5 | (4) | | | | |
| | After incubation: | 34.5 $\pm$ 0.6 | (4) | 29.9 $\pm$ 0.9 | (4) | - 4.7 | P > 0.10 |
| Krebs-Ringer bicarbunate buffer; 100 mM $Na^+$ replaced by $K^+$ | Before incubation: | 32.6 $\pm$ 1.6 | (4) | | | | |
| | After incubation: | 36.2 $\pm$ 1.7 | (4) | 15.5 $\pm$ 0.9 | (4) | + 29.3 | P < 0.001 |

# Sodium-Dependent Uptake of Iron-Transferrin in Rabbit Reticulocytes

W. C. Wise

Department of Physiology, Medical University of South Carolina, Charleston, South Carolina 29401, USA

(This investigation was supported in part by PHS Grant AMO 9069 and by NIH GRSG-RR 05420).

The influence of $Na^+$ on the transport of many solutes across biological membranes is of great interest. Na-dependent transport of sugars, amino acids and electrolytes is being investigated rather intensively. During this symposium, there has been discussion of the "Na-gradient hypothesis" and transport of non-electrolytes. This paper will discuss the $Na^+$ dependent uptake of iron, bound to its carrier protein transferrin. Some of the characteristics of this system indicate that the sodium gradient is a factor in the transport of iron by the maturing red blood cell.

Jandl and his co-workers (1959, 1963) reported that iron-saturated transferrin is preferentially bound to young red young blood cells with an affinity four to five times that of transferrin. This iron-saturated transferrin was suggested to be in equilibrium with unbound transferrin while the iron remained attached to the cell membrane. The first step in the transfer of iron from transferrin to the developing erythroid cell is the binding of the transferrin to the cell. Following the transferrin binding to the cell, the iron was believed to be released to move across the membrane, and then enter the heme synthesis pathway. More recent work of Morgan and Appleton (1969) has indicated that during the process of transferrin and iron-uptake, the transferrin molecules actually pass into the cells and are not exclusively localized to cell membrane receptors as reported by Jandl and Katz (1963). The present studies show that iron, from the iron-transferrin complex, is dependent on external sodium ions for part of its movement into the maturing red blood cell.

## Methods

Reticulocytosis was induced in rabbits with phenylhydrazine and blood was drawn by cardiac puncture with heparin as anticoagulant. No attempt was made to separate reticulocytes from mature erythrocytes which contributed less than 40 % to the population. It was demonstrated previously (Jandl, 1960) that the amount of iron and transferrin taken up by reticulocyte-rich blood is directly proportional to the reticulocyte count and mature erythrocytes do not take significant amounts of iron-transferrin. For simplicity, all blood cell suspensions are referred to as "reticulocytes". Cells at the outset were washed three times with isotonic tris-2-amino-2-(Hydroxymethyl)-1, 3-propanediol , brought to pH 7. 4 with HCl. Isotonicity of the tris-HCl was verified by microscopic examination of the cells. The standard medium contained 125 mM NaCl, 5 mM KCl, 1 mM $MgCl_2$, 19 mM Tris-HCl buffer, O. 5 ml rabbit serum, and $10^{-7}$M iron ($Fe^{59}Cl$). The iron and serum were mixed first to bind the iron to the serum transferrin. The amount of iron added to the serum was less than the iron-binding capacity of its transferrin and therefore the experiments follow the uptake by cells of the iron-transferrin complex. The experiments in which Na and K were omitted, isotonicity was maintained by the addition of appropriate amounts of tris-HCl or buffered choline-HCl. After washing, 0.02 ml of packed cells was introduced into each of a number of test tubes. Uptake was started by adding 5.0 ml of the standard or experimental media and the cells were incubated in a water bath at 37°C for the specific time intervals. Incubation was terminated by centrifugation for 1 minute. The supernate was aspirated and the cells washed twice. The supernate of successive washings was found to be free of $Fe^{59}$. Cells were hemolyzed in 2.0 ml of water and centrifuged at 1750 x g for three minutes. This force was sufficient to sediment the stroma completely from the rest of the hemolysate. Radioactivity of the supernate and stroma was counted in a well-type scintillation detector. Uptake was expressed as the percent of maximum counts in control sample for the longest incubation time in each experiment. Duplicate determinations were always made. Each experiment was compared with a control from the same population.

## Results

The uptake of iron-59 by rabbit reticulocytes in a medium with both sodium and potassium ions present is shown in Figure 1. The effects of the metabolic inhibitors

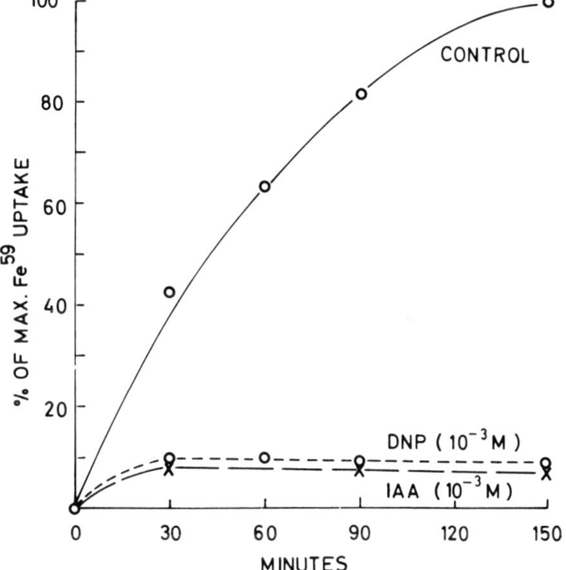


Fig. 1. Time course of the uptake of iron-59 by rabbit reticulocytes. Effect of 2, 4 dinitrophenol $(10^{-3}M)$ and iodoacetic acid $(10^{-3}M)$ on total cell iron-59 uptake. incubation was at 37°C. Cells were not pre-incubated with inhibitors

2,4 dinitrophenol (DNP) and iodoacetic acid (IAA) is also shown in this figure. The DNP and IAA were present initially and the cells were not pre-incubated with either inhibitors. The effects of these inhibitors on iron uptake was greater than 90% and was evident within 30 minutes incubation. The intracellular concentration of sodium ions did not change significantly until after 60 to 12o minutes of incubation.

In order to determine the effect of external sodium and potassium ions on iron-transferrin uptake, cells were incubated for 3 hours in a medium lacking both sodium and potassium ions. In Figure 2 the lack of sodium and potassium ions can be seen to reduce the accumulation of iron-59 in the supernant fraction of the hemolysate by at least 40%. The cells were not pre-incubated in the medium without sodium and potassium, so a reduction in the rate of iron-59 uptake by these cells is brought

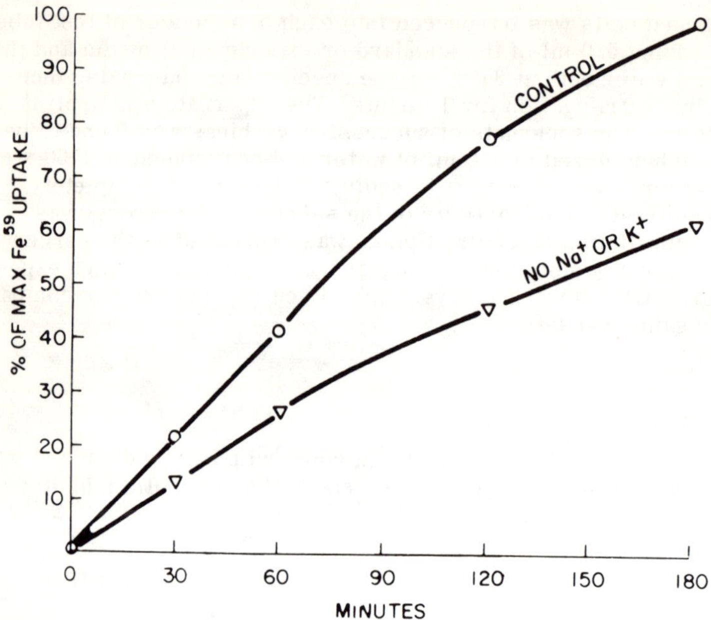



Fig. 2. Effect of absence of sodium and potassium ion on the time course of iron-59 uptake in the supernatant fraction of the hemolysate by rabbit reticulocytes. Isotinicity was maintained by the addition of the appropriate amount of tris-HCl.

about when sodium and potassium were excluded from the medium.

To determine the interaction of sodium and potassium on iron-transferrin uptake, experiments were performed in which sodium-ion concentration was varied and potassium-ion concentration was held constant. In Figure 3 the results of such

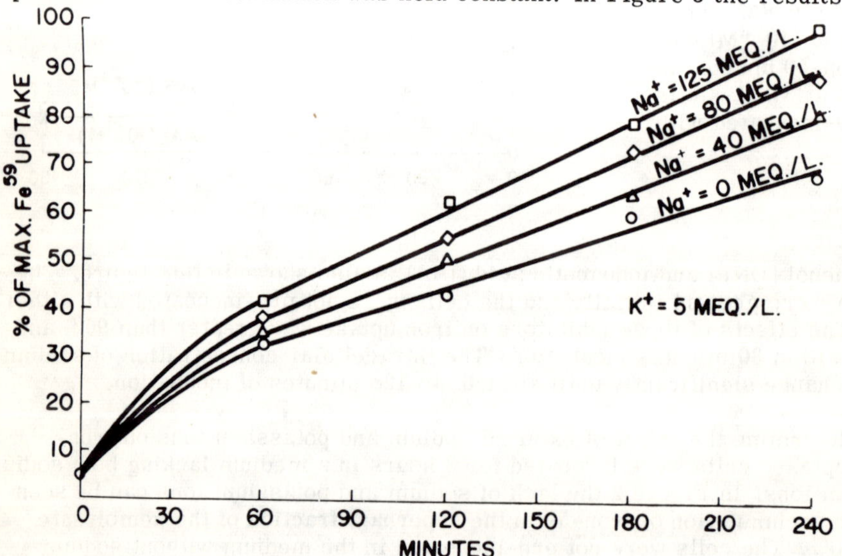



Fig. 3. Effect of different sodium ion concentrations on the time course of iron-59 uptake in the supernatant fraction of the hemolysate by rabbit reticulocytes. Isotonicity was maintained with tris-HCl or choline-HCl. Typical experiment

an experiment are presented. At a constant potassium-ion concentration of 5 mEq/l, the sodium-ion concentration was 0, 40, 80 and 125 mEq/l. Incubation was for 4 hours at 37°C and isotonicity was maintained with tris-HCl. Iron-59 accumulation in the supernatant fraction of the hemolysate was found to be reduced as the sodium-ion concentration was reduced.

The effect of increasing the sodium-ion concentration as the concentration of potassium ions is decreased on the rate of iron-59 uptake in the supernatant fraction is shown in Figure 4. Incubation was for 2 hours at 37°C. The rate of iron-59

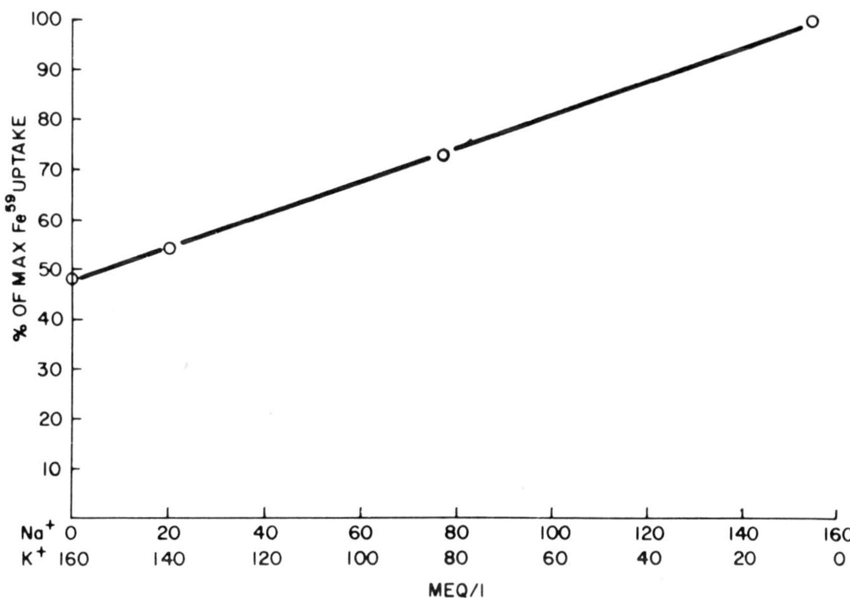


Fig. 4. Iron-59 uptake in the supernatant fraction of the hemolysate by rabbit reticulocytes after 120 minutes incubation at 37°C with increasing concentrations of sodium ion and decreasing potassium ion concentrations.

movement into the supernatant fraction of the hemolysate was directly proportional to increasing sodium-ion concentration and inversely proportional to potassium-ion concentration.

If the rate of iron uptake is determined at various concentrations of sodium in the absence of potassium,ions results of this type of experiment are shown in Figure 5. Sodium-ion concentration was increased from 0 to 140 mEq/l and incubation was for 4 hours at 37°C. When all sodium was replaced with tris-HCl, the rate of iron-59 uptake was reduced by 35%. As the sodium-ion concentration was increased from 0 to 40 mEq/l, no increase in the rate of uptake was found. But, at sodium-ion concentrations above 40 mEq/l, the rate of uptake was found to be directly proportional to the sodium-ion concentration.

Iron-59 uptake in rabbit reticulocytes with varying concentrations of iron in the fluid medium is shown for incubation times of 0-6 minutes with sodium ions present (Figure 6) and without sodium ions (Figure 7). Iron-transferrin was varied from $10^{-7}$M/l Fe to $10^{-8}$M/l Fe. The uptakes can be seen to be linear over this incubation period. Moreover, the effect of sodium ions is observed from the outset.

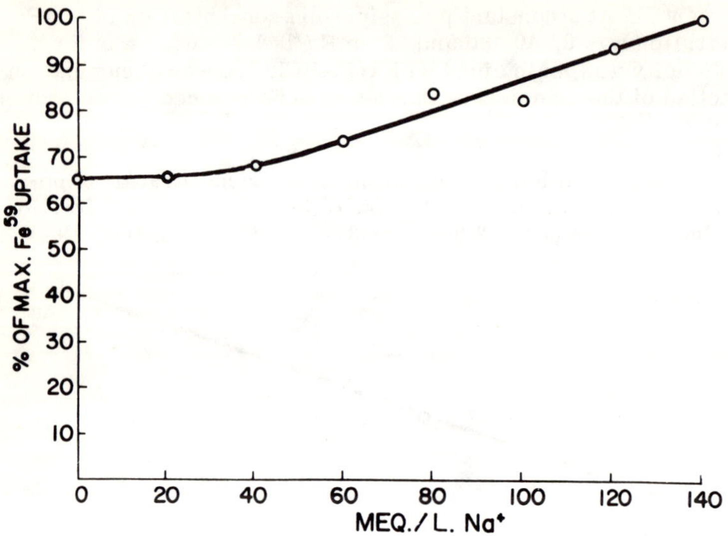

Fig. 5. Dependence of iron-59 uptake on extracellular sodium ion concentration. Potassium absent from incubation media. Isotonicity was maintained with tris-HCl

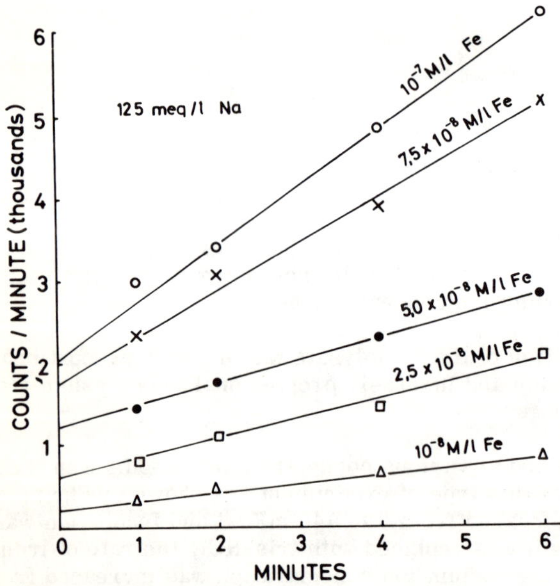


Fig. 6. Time course of iron-59 uptake by rabbit reticulocytes at different iron concentrations. Sodium ion at 125 mg/l present in incubation media

Figure 8 shows how the kinetic parameters for uptake were obtained from the results in Figures 6 and 7. The reciprocals of the uptake rates and media iron concentrations are shown as a Lineweaver-Burk plot. It should be pointed out that many assumptions are inherent in the use of Michaelis-Menten kinetics for such a multistage process. The apparent Km was $3 \times 10^{-5}$ moles/liter with sodium ions present and $5 \times 10^{-7}$ moles/l in the absence of sodium ions. The two lines have a

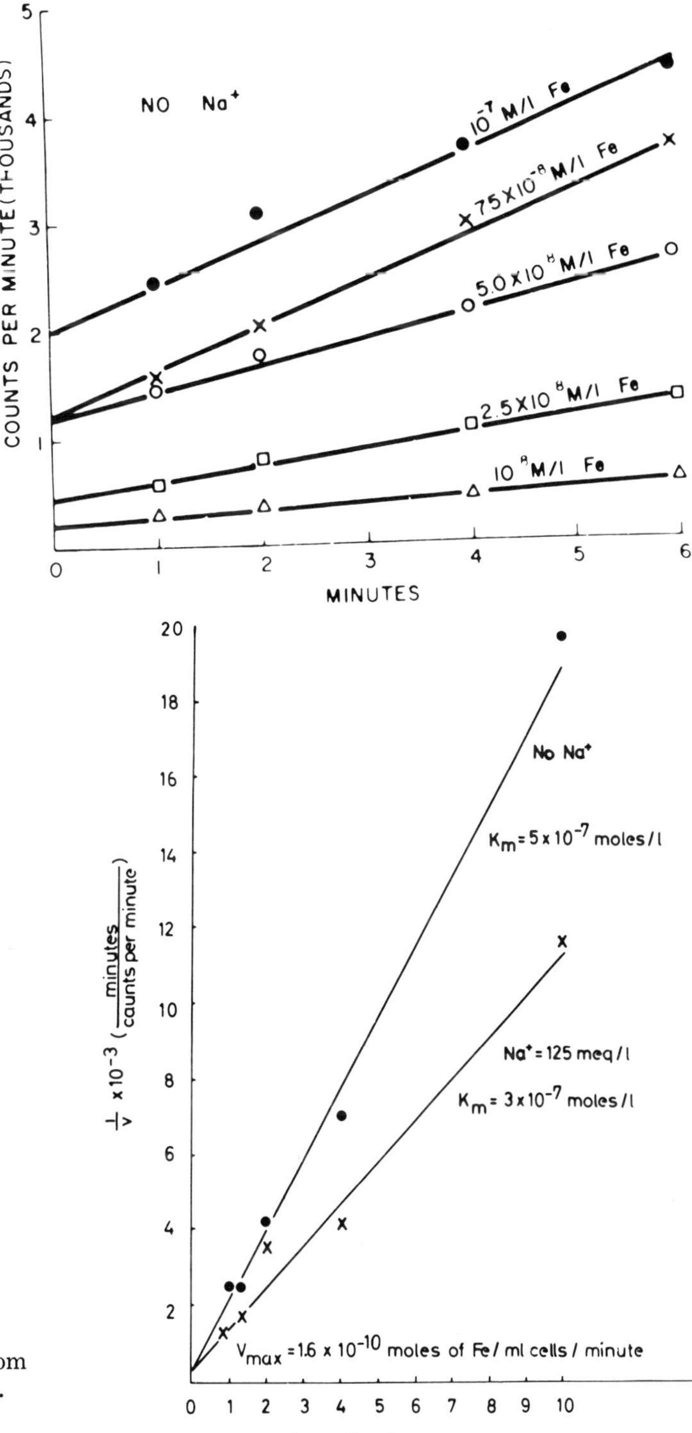



Fig. 7. Time course of iron-59 uptake by rabbit reticulocytes at different iron concentrations. Sodium was totally replaced by tris-HCl

Fig. 8. Estimation of kinetic parameters from results in Fig. 6 and 7. The apparent $K_m$ and $V_{max}$ value are shown

common intercept on the ordinate, indicating that the maximal uptake rate is the same in the presence and absence of sodium ions. At sodium-ion concentration intermediate to those shown, straight lines lying between those shown in Figure 8 are obtained.


## Discussion

Of the numerous studies of iron movement into maturing red blood cell, most have dealt with the transfer and attachment of transferrin to the developing erythroid membrane. Jandl et al. (1959) reported that only nucleated red cells and reticulocytes could accept the iron if it was bound to transferrin. Various observations indicate that these receptor sites are present only on the membranes of developing erythroid cells. From studies of Morgan et al. (1966) and others it seems likely that the transfer of iron takes place in three stages: the first a loose physical association with the receptor, the second a firmer bonding of the iron-transferrin complex, and the third, release of iron from the transport protein and the separation of the iron-binding protein from the cell. The mechanism whereby iron is released from its iron-binding protein is not clear. In more recent studies by Morgan and Baker (1969) and Baker and Morgan (1969) data has been presented showing that a significant proportion of the iron transferrin passes through the membrane and does not release the iron until the transferrin complex is intracellular.

Results presented here confirm the earlier work of Jandl and associates that the uptake of iron by the maturing red blood cell is dependent on metabolic energy. The uncoupling of oxidative phosphorylation by DNP and blocking of glycolysis with IAA both inhibited cellular uptake of iron. The onset of this inhibition of iron uptake was apparent during the first 30 minutes for both of these inhibitors.

We are now in a position to look at the effects of sodium ions alone on iron uptake. Figure 3 presents data to show that, at a constant potassium-ion concentration, the greater the sodium-ion concentration the greater the rate of iron uptake. By comparing the rates of uptake in Figure 4 to those in Figure 5, the presence of potassium at 160 mEq/l reduces the rate of uptake by some 15% more than if all the sodium was replaced by tris-HCl and no potassium was present in the medium. In Figure 5 the rate of uptake was the same at sodium concentration less than 40 mEq/l, but as the sodium-ion concentration is increased from 40 to 140 mEq/l, the rate of uptake increases proportionately to increasing sodium-ion concentration. In separate experiments under similar conditions, the internal concentration of sodium was found to be 30 mEq/l. Therefore, in these experiments, the external sodium concentration must be higher than the internal sodium concentration before the sodium-dependent uptake takes place. The ratio of sodium ions external to internal must be greater than one.

The absence of sodium and potassium ions in the incubation medium reduced the rate of iron uptake. Since evidence now exists that the iron-transferrin complex enters the cell before the iron is released, it could be concluded that sodium and/or potassium ions are necessary for the iron-transferrin complex to move into the cell. Evidence presented in Figures 4 and 5 along with work published earlier (Wise and Archdeacon, 1969) shows that the presence of potassium reduces iron uptake to a greater extent than the total absence of sodium. That is, potassium ions reduce iron uptake by maturing cells.

The question might be raised whether incubation in sodium – potassium–free media might alter the cell sodium and potassium concentration. Indirect evidence would indicate that this possibility would not be important in any explanation of the effect of the external concentration of ions on iron transport. First, in experiments in which rabbit reticulocytes were incubated at 4°C for 24 hours in sodium-potassium-free medium, the internal concentrations of both these ions change only a few mEq/l. Second, in experiments under similar conditions to those reported here with DNP and IAA, the internal sodium concentration did not significantly change until after 60 to 120 minutes incubation. Finally, in the uptake experiments which were for 6 minutes or less (Figures 6 and 7), the effects on iron uptake of incubation without sodium in the medium was evident within this incubation period. Intracellular ionic concentrations would not be abolished in this short incubation period.

We are now in a position to look at the kinetics of this sodium-coupled process. The uptake rate of the iron-transferrin complex can be described in terms of the "apparent Michaelis constant" and maximal uptake rate for various experimental conditions. The maximal uptake rates can be seen to be the same in the presence and the absence of sodium, as indicated by the two lines in Figure 8 having a common intercept. On the other hand, the $K_m$ is much greater in the absence of sodium than when sodium was present, since the slopes of the two lines are different. Similar kinetic patterns have been developed in studies of sugar transport (Crane, 1960) and alanine influx (Curran et al., 1967) in the intestine.

From these and other experimental data, a working hypothesis for iron uptake by the maturing red cell can be developed. Iron, bound to transferrin, attaches to the cell membrane. This membrane attachment is a function of sulfhydryl groups on the cell membrane (Wise, 1971). The iron-transferrin complex then moves through the membrane. The membrane process may derive its energy from at least two possible sources: one is presumably metabolic and inhibited by DNP and IAA; the other may come from energy inherent in the sodium gradient. The iron-transferrin complex gives up its iron intracellularly and the transferrin then moves out of the cell.

The inhibitory effect of potassium could be explained as causing a decrease in the affinity of the carrier for the iron-transferrin, even less than the carrier in the absence of sodium. This hypothesis is consistent with Crane's (1960) for sugar transport in the intestine.

A few comments are necessary about the iron uptake which is not sodium dependent. Our knowledge of this part of cellular uptake is quite meager. At this stage of the investigation, we can only say that energy is needed for this process.

## References


BAKER, E., MORGAN, E.H.: Biochemistry **8**, 295 (1969).
CRANE, R.K.: Physiol. Rev. **40**, 789 (1960).
CURRAN, P.F., SCHULTZ, S.G., CHEZ, R.A., FUISZ, R.E.: J. Gen. Physiol. **50**, 1261 (1967).
JANDL, J.H., INMAN, J.K., SIMMONS, R.L., ALLEN, D.W.: J. Clin. Invest. **38**, 161 (1959).
JANDL, J.W., KATZ, J.H.: J. Clin. Invest. **42**, 314 (1963).

MORGAN, E. H. , HUEHNS, E. R. , FINCH, C. A. : Am. J. Physiol. $\underline{210}$, 519 (1966).
MORGAN, E. H. , APPLETON, T. C. : Nature $\underline{223}$, 1371 (1969).
MORGAN, E. H. , BAKER, E. : Biochem. Biophys. Acta $\underline{184}$, 442 (1969).
WISE, W. C. , ARCHDEACON, J. W. : J. Gen. Physiol. $\overline{53}$, 487 (1969).
WISE, W. C. : Proc. Internat. Union Physiol. Sci. $\underline{9}$, 606 (1971).

# Carrier-Mediated, Na$^+$-Independent Translocation of Calcium across the Brush Border Membrane of Rat Duodenum in vitro

W. F. Caspary
Div. of Gastroenterology and Metabolism, Department of Medicine, University of Göttingen, Germany

Most of the work on the mechanism of calcium absorption by the intestine has been designed to the hypothesis of an active, vitamin D-dependent transport mechanism (10, 11, 15). Although at least two steps have been recognized in calcium transport across the intestinal wall in vitro (9), in most studies calcium transport has been treated as a single process measuring overall mucosal-serosal transfer. Different results for the mechanism of calcium transfer have been obtained: simple passive diffusion (4), facilitated diffusion (6, 13) and exchange diffusion (3, 14). Most authors, however, agree, that an active cation pump operating at the lateral-basal membrane of the mucosal epithelial cell is responsible for the rate-limiting steps of the over-all transport of calcium (15).

We were interested in the entry step mainly, trying to find out  whether this step might be rate-limiting,  secondly,  to test the entry step for energy dependence and Na$^+$ sensitivity as this holds for sugar (2) and amino-acid transport (12).

Entry of calcium was measured by incubating segments or sacs of everted rat duodenum in a phosphate-free medium with the appropriate concentrations of calcium and $^{45}$Ca as described earlier (1). Extracellular space was measured with $^3$H-D-mannitol. All results are corrected for extracellular space and are expressed as $\mu$ moles Ca accumulated/ml of tissue water or as:

$$\text{per cent filling} = 100 \cdot \frac{\mu \text{ moles/ml tissue water}}{\mu \text{ moles/ml medium}}$$

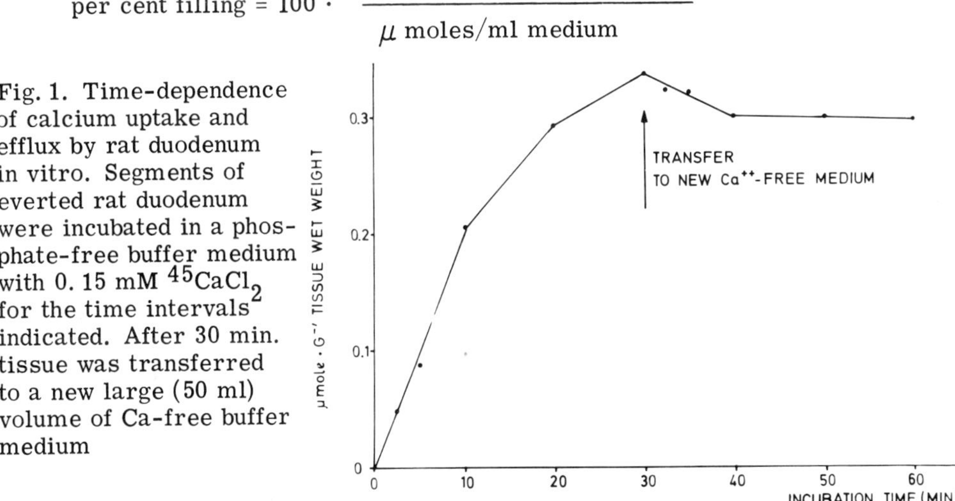



Fig. 1. Time-dependence of calcium uptake and efflux by rat duodenum in vitro. Segments of everted rat duodenum were incubated in a phosphate-free buffer medium with 0. 15 mM $^{45}$CaCl$_2$ for the time intervals indicated. After 30 min. tissue was transferred to a new large (50 ml) volume of Ca-free buffer medium

Uptake of $^{45}$Ca was linear up to 10 minutes of incubation (Fig. 1). Further decrease of calcium influx/unit time may then be due to efflux, but as shown, too, in fig. 1, efflux was much slower than influx if tissue was transferred to a new large volume of buffer without calcium in the incubating medium. In kinetic experiments, therefore, an incubation time of 10 minutes was chosen to represent initial rates of uptake.

Measuring the concentration-dependent uptake of calcium over a wide range of calcium concentrations ($10^{-4}$M - $5 \cdot 10^{-2}$M) we found (Fig. 2) that at the lower concentrations calcium uptake was a saturable process, but at the higher concentrations (above 5 mM) uptake rates were linear with increasing concentrations of calcium in the medium. If these results are plotted according to Lineweaver-Burk,

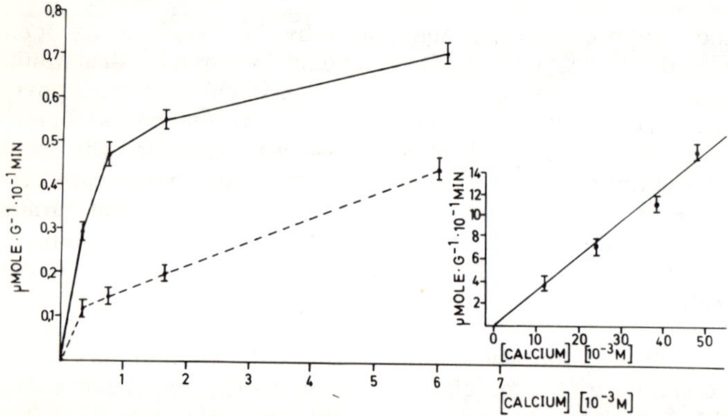

Fig. 2. Concentration-dependent uptake of calcium by segments of everted rat duodenum in vitro in presence ( - - - - - -) and absence ( ___________ ) of 2, 4-dinitrophenol (5 x $10^{-4}$M). Results are given in means ± S. E. M. (n = 6)

calcium influx exhibits an apparent transport Km of 0.9 mM at the lower concentrations, consisting with translocation by a carrier-mediated process. These results agree with those in the literature taking mucosal-serosal transfer for velocity (6, 13). Entry at high medium concentrations of calcium, however, is diffusional (Fig. 3).

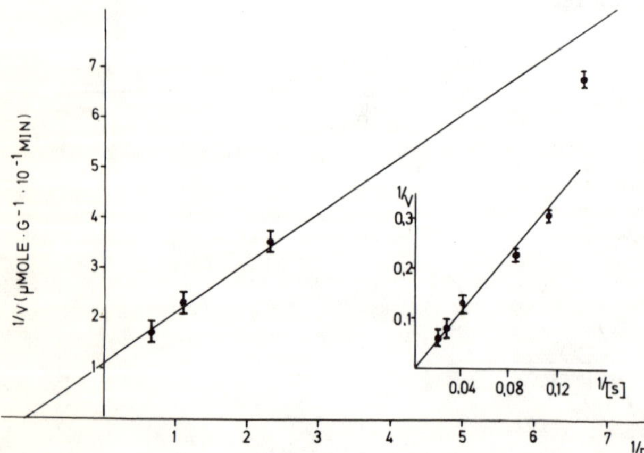

Fig. 3. Lineweaver-Burk plot of the concentration-dependent uptake of calcium into segments of everted rat duodenum in vitro. Incubation was for 10 min. in a phosphate-free medium. Results are given in means + S. E. M (n = 6). Ordinate = $1/\mu$ moles $\cdot$ G$^{-1}$ $\cdot$ $10^{-1}$ min, abscissa = $1/10^{-3}$M Ca

Overall mucosal-serosal transport of calcium has been found to be Na$^+$-sensitive (7). Influx of calcium, however, was not Na$^+$-sensitive if Na$^+$ was replaced by Tris, choline or mannitol. In contrary, uptake of calcium was increased if Na$^+$

was replaced by D-mannitol, a finding we cannot readily explain at the moment.
Recent findings of Rose and Schultz (8) show that the cell interior of rabbit small
intestine was negative and became even more negative if $Na^+$ was replaced by other
cations like Tris and choline. This potential difference could account already for
an eight-fold accumulation of a divalent cation without the necessity of invoking
active transport processes. Ouabain did also have no effect on initial uptake rates,
but 2, 4-dinitrophenol had an inhibitory effect (Fig. 2, 4).

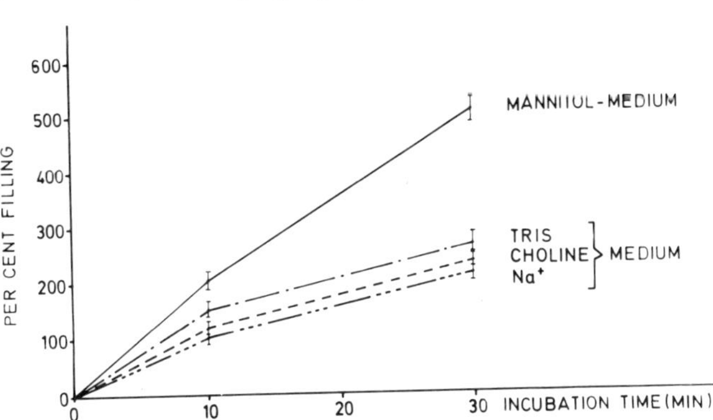


Fig. 4. Effect of $Na^+$ replacement by D-mannitol, choline or Tris on tissue calcium accumulation in segments of rat duodenum. Substrate: 0. 15 mM $CaCl_2$. Incubation was carried out for the time indicated in a phosphate-free buffer medium. Volume: 10 ml

Thus the carrier-mediated entry-step for calcium at the lower calcium concentrations seems to be $Na^+$-insensitive, but energy-dependent.

Using the everted-sac preparation and incubating for 60 minutes (Fig. 5) we found

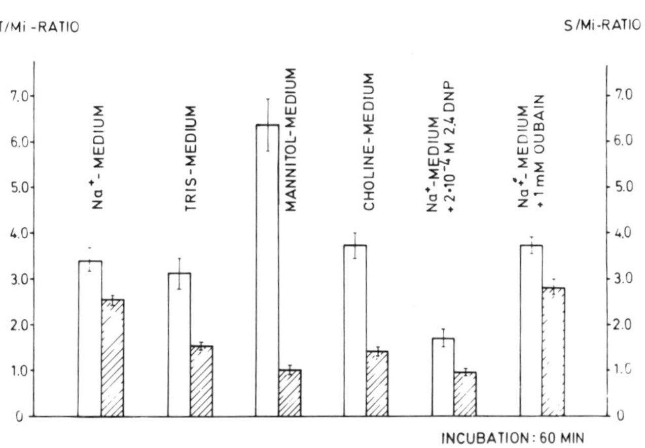


Fig. 5. Effect of $Na^+$ replacement by D-mannitol, choline or Tris and addition of 2, 4-dinitrophenol or ouabain on tissue accumulation and mucosal-serosal transfer of calcium in everted sacs of rat duodenum. Sacs were incubated for 60 min. in a phosphate-free medium with total $Na^+$ replaced by D-mannitol, choline or Tris. $^{45}Ca$ was analyzed in the tissue and in the serosal compartment. Results for tissue accumulation are expressed in T/Mi-ratios (tissue/initial mucosal medium concentration), serosal accumulation is expressed in S/Mi-ratios (serosal/initial mucosal medium concentration). Results are means $\pm$ S. E. M (n = 5). Open bars: tissue accumulation (T/Mi), hatched bars: serosal accumulation (S/Mi)

that tissue calcium accumulation was only reduced if 2, 4-dinitrophenol was present, whereas $Na^+$ replacement by Tris, choline, mannitol or presence of ouabain did not affect the tissue accumulation of calcium. In contrary, however, overall mucosal-serosal transfer was markedly depressed by $Na^+$ replacement. Ouabain did not affect mucosal-serosal transfer of calcium if present in the mucosal and serosal compartment. The ineffectiveness of ouabain may be explained by the relative low sensitivity of rat intestinal $Na^+$-$K^+$-ATPase to cardiac glycosides (5). Thus, overall mucosal-serosal transport of calcium in contrary to the carrier-mediated influx step is $Na^+$-sensitive and energy-dependent.

Calcium entry into the mucosal epithelial cells, as characterized by the saturable, $Na^+$-insensitive component will predominate at lower concentrations. We know that the intracellular electrical potential in rabbit ileum is negative (8). If one assumes this to be true for the rat duodenum, too, entry of calcium across the brush border membrane is down an electrochemical gradient which could account already for a several-fold accumulation of calcium in the mucosal epithelial cells. Tissue poisoning with KCN or iodoacetate decreases markedly the intracellular negativity of the electrical potential (8) thus decreasing the favouring electrochemical gradient for calcium influx. Transport across the basal or lateral border of the mucosal epithelial cells is then up an electrical gradient and occurs by a $Na^+$-dependent and energy-dependent mechanism.

## Acknowledgements

The author thanks Miss A. Thinius and Miss H. Römhild for skilful technical assistance. The work was supported by a grant of the Deutsche Forschungsgemeinschaft (Ca 71/1).

## References


1. CASPARY, W. F. , STEVENSON, N. R. , CRANE, R. K. :  Evidence for an intermediate step in carrier-mediated sugar translocation across the brush border membrane of hamster small intestine. Biochim. Biophys. Acta 193, 168-178 (1969).
2. CRANE, R. K. : Absorption of sugars. In: Handbook of Physiology, Section 6: Alimentary Canal, Vol. III, C. F. Code (editor), American Physiological Society, Washington, D. C. 1968, p. 1323 - 1351.
3. DUMONT, P. A. , CURRAN, P. F. , SOLOMON, A. K. : Calcium and strontium in rat small intestine. Their fluxes and their effects on $Na^+$-flux. J. General Physiol. 43, 1119-1136 (1960).
4. HELBOCK, H. J. , FORTE, J. G. , SALTMAN, P. : The mechanism of calcium transport by rat intestine. Biochim. Biophys. Acta 126, 81-93 (1966).
5. LEOPOLD, G. , FURUKAWA, E. , FORTH, W. , RUMMEL, W. : Binding of cardiac glycosides to isolated jejunal brush borders from rat and guinea pig and their influence on membrane phosphatase system. Biochem. Pharmacol. 20, 1109-1117 (1971).
6. MARTIN, D. L. , DeLUCA, H. F. : Calcium transport and the role of vitamin D. Archives of Biochemistry and Biophysics 134, 139-148 (1969).
7. MARTIN, D. L. , DeLUCA, H. F. : Influence of sodium on calcium transport by rat small intestine. Am. J. of Physiology 216, 1351-1359 (1969).
8. ROSE, R. C. , SCHULTZ, S. G. : Studies on electrical potential profile across rabbit ileum. Effects of sugars and amino acids on transmural and transmucosal electrical potential differences. J. Gen. Physiology 57, 639-663 (1971).

9. SCHACHTER, D., KOWARSKI, S., FINKELSTEIN, J.D., WAND MA, R.T.: Tissue concentration differences during active transport of calcium by intestine. Am. J. of Physiol. 211, 1131 - 1136 (1966).
10. SCHACHTER, D., KIMBERG, D.V., SCHENKER, H.: Active transport of calcium by intestine: Action and bioassay of vitamin D. Am. J. of Physiol. 200, 1263-1271 (1961).
11. SCHACHTER, D., ROSEN, S.M.: Active transport of $^{45}$Ca by the small intestine and its dependence on vitamin D. Am. J. of Physiol. 196, 357-362 (1959).
12. SCHULTZ, S.G., CURRAN, P.F.: Coupling of sodium and organic solutes. Physiological Reviews 50, 637-718 (1970).
13. WALLING, M.W., ROTHMAN, S.S.: Apparent increase in carrier-affinity for intestinal calcium transport following dietary calcium restriction. J. Biolog. Chem. 245, 5007-5011 (1970).
14. WASSERMAN, R.H., KALLFELZ, F.A.: Vitamin $D_3$ and unidirectional calcium fluxes across the rachitic chick duodenum. Am. J. Physiology 203 221-224 (1962).
15. WASSERMAN, R.H.: Calcium transport by the intestine: a model and comment on vitamin D action. Calcified Tissue Research 2, 301-313 (1968).